LYSOSOMES - ASSOCIATED DISEASES AND METHODS TO STUDY THEIR FUNCTION

Edited by **Pooja Dhiman Sharma**

Lysosomes - Associated Diseases and Methods to Study Their Function
http://dx.doi.org/10.5772/65224
Edited by Pooja Dhiman Sharma

Contributors

Barbara Hissa, Luciana O Andrade, Eleonore Fröhlich, Haruo Shimazaki, Sylwia Chojnowska, Alina Kępka, Slawomir Dariusz Szajda, Napoleon Waszkiewicz, Krzysztof Zwierz, Rubén G. Contreras, Catalina Flores-Maldonado, Ruth Rincón-Heredia, Jessica Paulina Campos-Blázquez, Vicky García Hernández, Odette Verdejo-Torres, Agustin Ruiz-Cabrera, Marcelo Nociari, Enrique Rodriguez-Boulan, Fenghua Hu, Peter Sullivan, Xiaolai Zhou

Published by InTech

Janeza Trdine 9, 51000 Rijeka, Croatia

Publishing Process Manager Dajana Pemac
Technical Editor SPi Global
Cover InTech Design team

First published August, 2017
Printed in Croatia
Legal deposit, Croatia: National and University Library in Zagreb

Additional hard copies can be obtained from orders@intechopen.com

Lysosomes - Associated Diseases and Methods to Study Their Function, Edited by Pooja Dhiman Sharma
p. cm.
Print ISBN 978-953-51-3507-4
Online ISBN 978-953-51-3508-1

For EU product safety concerns :
IN TECH d.o.o., Prolaz Marije Krucifikse Kozulić 3, 51000 Rijeka, Croatia, info@intechopen.com or visit our website at intechopen.com.

Meet the editor

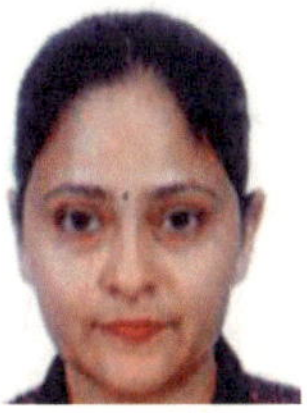

Dr. Pooja Dhiman obtained an MBBS degree followed by an MD degree from the Postgraduate Institute of Medical Education and Research (PGIMER), Chandigarh, India. After completing three years of senior residency, she qualified for a Membership in the Indian Board of Clinical Biochemistry (MIBCB). Currently, she is pursuing a PhD degree in Medical Biochemistry at the Jawaharlal Institute of Postgraduate Medical Education and Research (JIPMER), Puducherry, India. Presently, she has 18 publications in recognized international journals and a book chapter.

Contents

Preface

This book includes scholarly contributions by various experts in the field of lysosomal diseases. Each contribution comes as a separate chapter complete in itself but directly related to the book's topics and objectives. The target audience comprises scholars and specialists in the field.

The book is divided into two sections: "Lysosomes and Diseases" and "Techniques - Lysosomal Function."

The first section includes five chapters: "Lipofuscin Accumulation into and Clearance from Retinal Pigment Epithelium Lysosomes: Physiopathology and Emerging Therapeutics," "Lysosomal Dysfunctions in Hereditary Spastic," "Lysosomes: How Plasma Membrane Repair Route Can Be Hijacked by Parasites?," "Autophagy-Lysosomal Dysfunction in Amyotrophic Lateral Sclerosis and Frontotemporal Lobar Degeneration," and "Lysosomal Degradation of Junctional Proteins."

The second section includes two chapters: "Cellular Screening Methods for the Study of Nanoparticle-Induced Lysosomal Damage" and "Diagnostic Application of Lysosomal Exoglycosidases."

We hope that the book will be a useful tool for all those interested in this field.

Pooja Dhiman Sharma
Department of Biochemistry
JIPMER
Dhanvantri Nagar
Gorimedu, Puducherry
India

Section 1

Lysosomes and Diseases

Chapter 1

Lipofuscin Accumulation into and Clearance from Retinal Pigment Epithelium Lysosomes: Physiopathology and Emerging Therapeutics

Marcelo M. Nociari, Szilard Kiss and
Enrique Rodriguez-Boulan

Additional information is available at the end of the chapter

http://dx.doi.org/10.5772/intechopen.69304

Abstract

Photoreceptors undergo a constant renewal of their light sensitive outer segments (POSs). In the renewal process, 10% of the POS mass is daily phagocytized by the adjacent retinal pigment epithelium (RPE). POS contain vast amounts of 11-cis retinal and all-trans-retinal, two highly reactive vitamin A aldehydes that spontaneously dimerize into lipid bisretinoids (LBs) and accumulate into RPE lysosomes during phagocytosis. As LBs are refractory to lysosomal hydrolases and RPE cells do not divide, this accumulation is irreversible and results in the formation of lipofuscin granules. Lipofuscin accumulation is toxic for RPE cells through a variety of light-dependent and light-independent mechanisms. Beyond a threshold, RPE cells die resulting in secondary loss of overlying photoreceptors. Currently, there are no effective treatments for retinal disorders associated with genetic or age-associated LB accumulation, such as Stargardt disease and age-related macular degeneration (AMD). Thus, there is a great need for medical interventions. Here, we discuss the current understanding of lipofuscin's pathogenicity and the status of different strategies under development to promote LB elimination from RPE lysosomes.

Keywords: lipofuscin, Stargardt, age-related macular degeneration (AMD), bisretinoids, retinal pigment epithelium (RPE), cyclodextrins, cellular clearance, TFEB, lysosome

1. Introduction

To understand the origin and consequences of the lysosomal accumulation of lipofuscin in the eye, a basic knowledge of retinal function and organization is required.

1.1. The retinal pigment epithelium (RPE) in vertebrate's eyes

Light entering the eye gets refracted by the cornea and lens on the neural retina, where photoreceptors (PR) convert photons into a cascade of chemical and electrical events that propagate to second-order (horizontal, bipolar, and amacrine cells) and third-order (ganglion cells) retinal neurons, which distribute this information to various visual centers of the brain through the fibers of the optic nerve. The bodies of PR cells, rods and cones, display three sectors (**Figure 1**): the *outer segment*, filled with stacks of disks densely packed with light-sensitive photopigment; the *inner segment*, filled with genetic, biosynthetic, and metabolic organelles

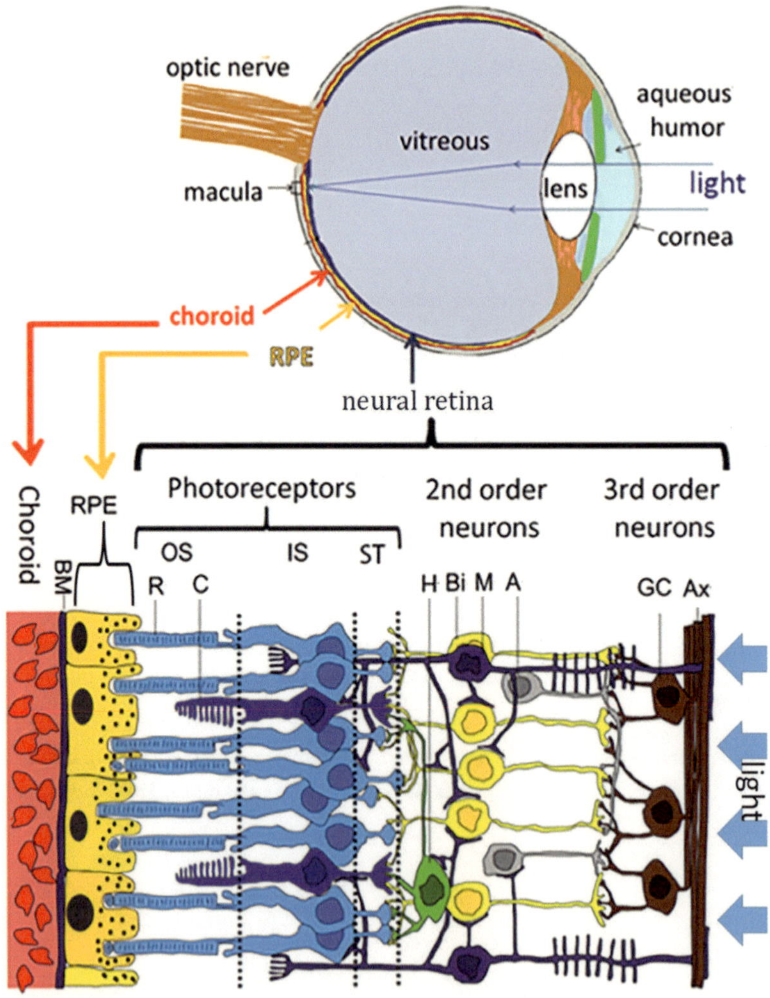

Figure 1. Basic structure of retina's inverted architecture. Light travels through ganglion cells (GC) as well as second-order neurons comprising bipolar (Bi), amacrine (A), horizontal (H), and Muller (M) cells, the retina. RPE localizes between the choroid capillaries and the cone (C) and rod (R) photoreceptors. Photoreceptors outer segments (OS), inner segments (IN), and synaptic terminals (ST).

(nucleus, endoplasmic reticulum, Golgi complex, ribosomes, mitochondria); and the *synaptic terminal* that connects with bipolar neurons of the retina. In vertebrates, the retina is inverted in the sense that light passes through secondary and tertiary neuronal layers in the inner retina before reaching the rods and cones in the outer retina (**Figure 1**). Photoreceptors are metabolically very active cells that require large amounts of nutrients and generate massive amounts of waste. Provision of nutrients and removal of waste are critical support tasks performed by the retinal pigment epithelium (RPE), a monolayer of cuboidal cells tightly opposed to the photoreceptors through a narrow *subretinal space*. The multifunctional RPE cells constitute the blood-outer retina barrier that controls the movement of nutrients, ions, water, gases, and wastes between the photoreceptors and underlying choroidal blood vessels [1], absorbs excess light through their melanin granules, performs segments of the visual cycle that regenerates the visual pigment, performs daily phagocytosis of photoreceptor outer segments, and produces trophic factors necessary for photoreceptor survival [2]. It may be rationalized that a fundamental objective of the inverted retina design is to bring photoreceptors in close contact with RPE, a key relationship for the integrity of the retina.

1.2. Role of RPE in visual-pigment regeneration

In 1967, George Wald was awarded the Nobel Prize for revealing the essential role of Vitamin A for vision [3]. Vitamin A entirely is derived from the diet. A critical function of RPE cells is to supply the vitamin A-derived chromophore, 11-cis retinal, required for the light-sensing function of visual pigments to photoreceptor cells. Visual pigments are G-protein receptors (opsins) covalently linked to 11-cis-retinal. Opsins cannot respond to light by themselves but need their prosthetic group, 11-cis-retinal that undergoes cis-trans isomerization upon illumination. The absorption characteristics of 11-cis retinals, in different pigments, are determined by the opsins. Humans have four types of visual pigments: *rhodopsin*, expressed by rod photoreceptors, which is sensitive to dim light and provides black-and-white vision; *L-opsin*, *M-opsin*, and *S-opsin* expressed by cone photoreceptors sensitive to red, green, and blue color lights, respectively [4]. When light strikes the visual pigments, it promotes isomerization of 11-cis retinal into all-trans retinal (ATR), which in turn promotes the structural rearrangement of opsin into an active conformation that initiates phototransduction (**Figure 2**).

To recover photosensitivity, opsin must be regenerated by releasing all-trans retinal and binding new 11-cis retinal. Released all-trans retinal is pumped out of the disks into the cytosol by a photoreceptor specific ATP-binding transporter (ABCA4) and reduced to all-trans-retinol by all-trans-retinal dehydrogenases (RDH8 and RDH12). All-trans-retinol diffuses into the RPE where it is esterified by lecithin:retinol acyltransferase (LRAT) to all-trans-retinyl esters, which are stored in retinosomes. All-trans-retinyl esters are isomerized by 65-kDa RPE-specific protein (RPE65) to 11-cis retinol, which is oxidized to 11-cis retinal before traveling back to the photoreceptors outer segment where it is again conjugated to an opsin to form new, functional visual pigment. These series of metabolic steps, by which all-trans-retinal is converted to regenerate the visual pigments, constitute the classical visual cycle [5]. As described above, the RPE performs a number of critical steps in the classical visual cycle that provides 11-cis retinal to rods and cones. There is also a cone-specific visual cycle [6], which

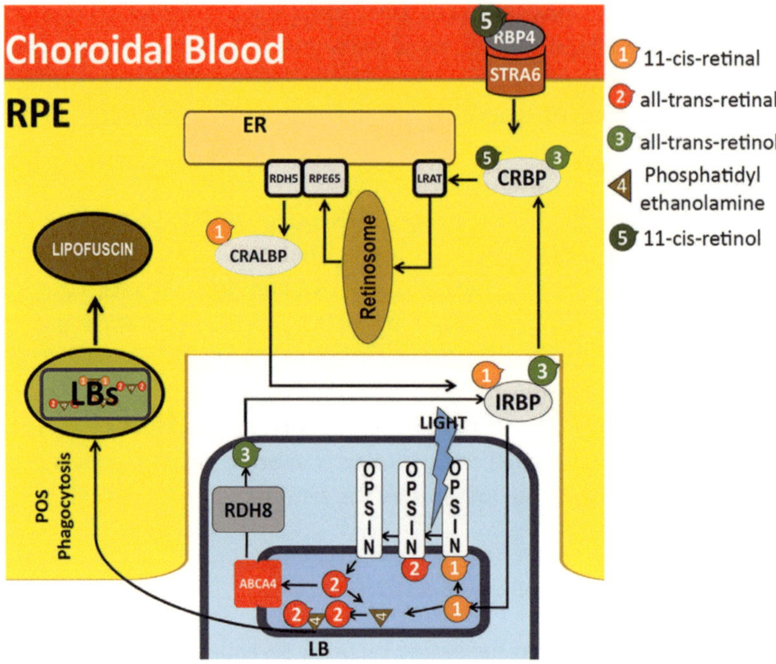

Figure 2. Classical visual cycle. Light induces cis-trans isomerization of retinal in visual pigments. All-trans retinal is released from opsins and pumped to the cytosol by ABCA4 where is reduced to all-trans-retinol by dehydrogenases (RDH8). All-trans-retinol diffuses into RPE via a series of Interphotoreceptor (IRBP) and cellular (CRBP and CRALBP) retinoid-binding proteins. In RPE all-trans-retinol is esterified by lecithin:retinol acyltransferase (LRAT) and is sored as retinyl esterin lipid droplets called retinosomes. When mobilized, all-trans-retinyl esters are isomerized to 11-cis-retinol by RPE65 and oxidized to 11-cis retinal by RDH5 before traveling back to the photoreceptor disks where it is again conjugated to an opsin to form new, functional visual pigment. Circulating retinol-binding protein 4 (RBP4) and its membrane receptor STRA6 coordinate 11-cis-retinol uptake into cells with high demand of Vitamin-A.

is RPE independent, that will not be discussed here. The RPE can also generate 11-cis retinal from vitamin A captured via its transmembrane transporter Stra6 from the choroidal circulation. Another important source of 11-cis retinal is the retinosomes, i.e., intracellular lipid droplet deposits of all-trans-retinyl esters in the cytoplasm of RPE cells (**Figure 2**).

1.3. Lipid-bisretinoid (LB) biogenesis

Vitamin A aldehydes (retinaldehydes) are highly reactive molecules capable of forming adducts with biological amines without the need for a catalyst [7]. In the disks of photoreceptor outer segments (POSs), retinaldehyde concentrations are relatively high, due to the *all-trans-retinal* released by photo-transduction and the *11-cis-retinal* conveyed for the regeneration of visual pigments (**Figure 2**). Not surprisingly, POS' retinaldehydes tend to covalently react with the amine group of phosphatidyl-ethanolamine (PE) to form N-retinylidene-PE (NRPE), which reacts with the second molecule of retinal to produce *lipid-bisretinoids* (*LBs*). Thus, LBs are a family of adducts, all structurally related, that derive from the condensation of two retinaldehydes with one PE molecule [8, 9].

1.4. Photoreceptor renewal

Because of their task in vision and proximity to the fast flowing choroidal capillaries, photoreceptors are continually exposed to high doses of radiant energy and oxygen, which makes them prone to photo-oxidative damage. To secure long (decades) of useful life, under these demanding conditions, photoreceptors undergo a daily renewal process wherein the most distal tips of their POS, comprising the ~100 oldest disks, are removed and equivalent number are basally produced to maintain constant outer segment length [10]. This cellular renewal process has a circadian rhythm. The rods shed POS most vigorously in the morning, whereas cones shed more vigorously at the onset of darkness [11]. The enormous amount of waste daily generated by this process is cleared by the adjacent RPE cells. In the mammalian eye, one RPE cell serves approximately 40 photoreceptor cells, each of which sheds ~7% of its mass per day. RPE engulfs and degrades POS fragments via a receptor-mediated phagocytic process similar to that involved in macrophage-mediated removal of apoptotic cells [12, 13]. This is an impressive metabolic task for RPE, since each cell must ingest and digest ~4000 disks before the next phagocytic load. Thus, RPE is one of the most active phagocytic cells in the body. Because RPE cells do not divide, they must completely dispose this daily material to avoid POS components buildup in their lysosomes.

1.5. RPE lipofuscin accumulation

"Lipofuscin" is the generic name given to subcellular material that accumulates with age within the lysosomal compartment of a variety of postmitotic cells and is characterized by its golden-orange autofluorescent emission. Very few compounds of animal origin exhibit fluorescent emissions in the lipofuscin's region of the spectrum [14]. A fairly rigid structure with highly conjugated double bond system is necessary for such fluorescence because, accumulation of lipofuscin is considered an universal biomarker of aging, as it is also referred to as "age pigment". Lipofuscins are resistant to degradation by lysosomes, proteasomes, and are not evidently exocytosed. Hence, their accumulation appears irreversible in cells that do not divide. Most lipofuscins stain positive for proteins, lipids, and carbohydrates [15]. Their exact composition varies among tissues but most commonly contains a large proportion of incompletely degraded proteins [16]. The RPE is one of the tissues with the largest buildups of lipofuscin. RPE lipofuscin increases with age in all healthy eyes [17, 18]. It localizes in lysosomal bodies of the RPE [19] and can occupy ~ 20% of the cytoplasmic space by 80 years of age [20].

In order to illuminate the cellular processes responsible for the formation of RPE lipofuscin, several groups attempted to analyze RPE-lipofuscin's chemical composition. Eldred and Katz [21] were the first to isolate the fluorescent pigments of the RPE lipofuscin. Spectroscopy and mass spectroscopy analyses of lyophilized chloroform extracts of RPE cells from healthy donors of different ages revealed that the most common fluorophore was a lipid-bisretinoid [22], N-retinylidene-N-retinylethanolamine also called A2E [23]. Protocols for *in vitro* synthesis of A2E as well as its incorporation into lysosomes of cultured RPE were developed, allowing to model RPE lipofuscin accumulation *in vitro* [24, 25]. Further efforts to isolate and characterize the remaining chromophores in the chloroform extracts from RPE lipofuscin granules, yielded additional LBs, including A2-GPE (A2-glycero-phospho-ethanolamine),

A2-DHP-PE (A2-dihydropyridine-phosphatidyl-ethanolamine), all-trans-retinal dimer (ATRD), and all-trans-retinal dimer phosphatidyl-ethanolamine (ATRD-PE) (**Figure 3**) as well as several higher molecular weight hydrophobic polymers derived from the reaction between A2E and its oxidation products [26–29].

Of note, the chloroform-insoluble fraction of the RPE lipofuscin, which represents 70% of its dry weight, was not analyzed in these studies. Thus, to fully characterize this fraction, Schutt et al. performed a proteome analysis of sucrose-purified RPE granules [30]. They identified 65 abundant cellular proteins, which included structural, metabolic, mitochondrial, chaperone, transmembrane, and signaling transduction proteins. Many of these proteins were modified by reactive carbonyl compounds (4-hydroxynonenol and malonyldialdehyde) and exhibited advanced glycation end products (AGEs) [31]. A second study by Warburton et al. [32] identified 41 proteins, most of which included phagosomal, lysosomal, and photoreceptor proteins (including rhodopsin) in agreement with the notion that RPE lipofuscin was mainly a buildup of undigested POS material. Surprisingly, only 12 proteins (11%) of Warburton's list were common with the 65 proteins identified by Schutt et al. This discrepancy probably results from variations in the purity of sucrose-isolated granules [33] and from the fact that lipofuscin proteins are microheterogeneous in size due to abundant oxidative modifications while contaminant proteins are intact and therefore, run as well-focused spots. In a third study, Ng et al. [34] analyzed the composition of highly purified RPE-lipofuscin granules devoid of

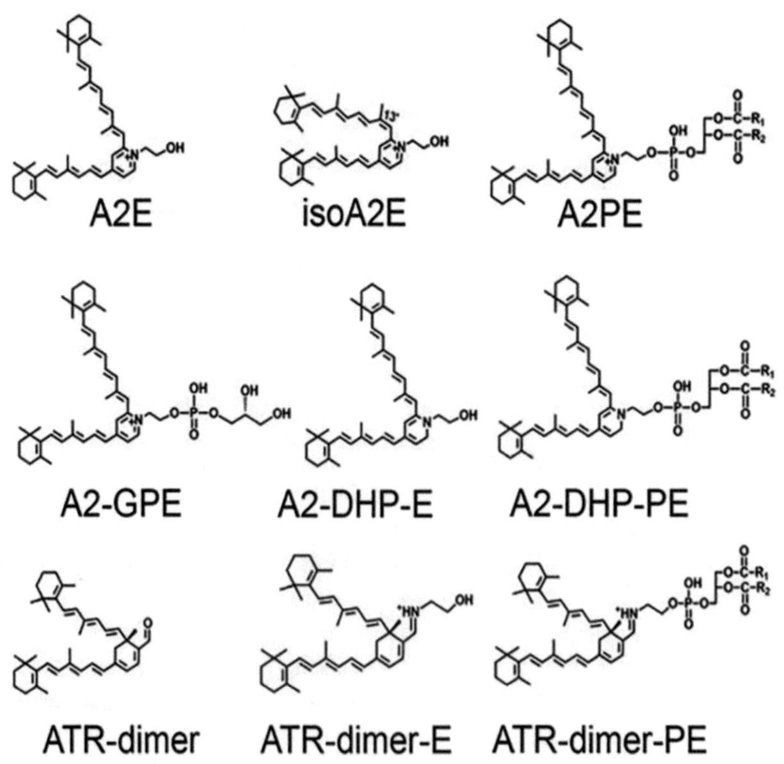

Figure 3. LBs found in RPE lipofuscin. R1, R2 are fatty acids with 14 to 22 carbons and 0 to 6 double bonds.

membranes and reported that the luminal material was 98% lipids, mostly LBs [35]. Taking all this information into account, the current concept is that RPE lipofuscin originates from LBs in photoreceptors and is transferred to RPE lysosomes during POS phagocytosis. This model is supported by animal studies that show that accumulation of lipofuscin in the RPE only occurs if (i) there is a supply of 11-cis-retinal to synthesize visual pigments, as RPE65$^{-/-}$ mice display no lipofuscin [36] and (ii) there is phagocytosis of POS, since no accumulation of lipofuscin is detectable in phagocytosis-defective animals. In healthy individuals, LB formation occurs slowly because the concentrations of retinaldehydes are relatively low, thereby, taking many years to generate significant amounts of LBs. In contrast, in individuals with mutations in ABCA4, the formation of LBs is dramatically accelerated [37].

1.6. Cellular toxicity caused by RPE lipofuscin accumulation

In retinal diseases associated with the accumulation of LBs in RPE lysosomes, vision loss is the result of the death of photoreceptor cells secondary to the functional impairment of RPE. Cell culture experiments have shown that lysosomal accumulation of LBs can cause RPE cell death [38]. However, how exactly lipofuscin accumulation disrupts RPE performance and viability is not fully understood. The variety of LB-elicited toxic mechanisms proposed so far (**Figure 4**) and their investigation as potential pharmacological targets are discussed below.

1.6.1. Phototoxicity

In vitro data with both, whole lipofuscin granules [39] or individual bisretinoids (A2E [40], all-trans-retinal dimer [27] and A2-GPE [41]) loaded into lysosomes have shown that LBs

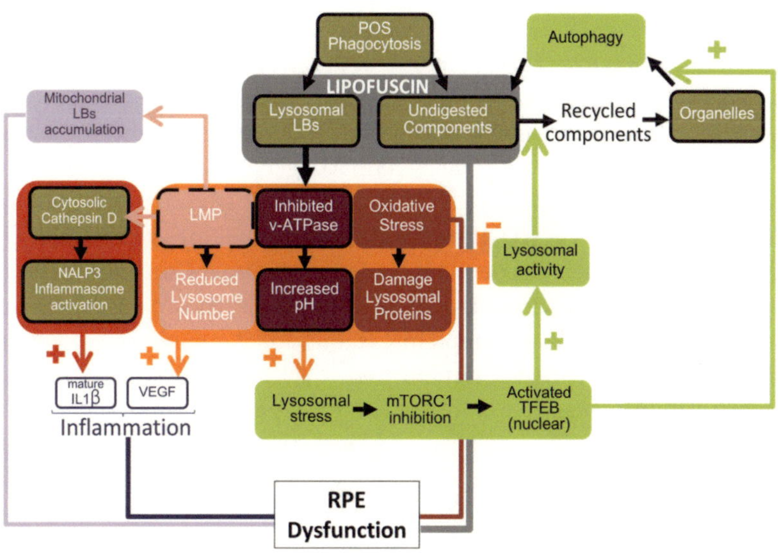

Figure 4. Modeling the effect of LB accumulation on RPE's lysosomal function. LBs cause loss of lysosomal function by inhibition of v-ATPase, oxidative damage, and/or LMP leading to buildup of undigested POS, autophagy material and TFEB-mediated lysosomal stress. TFEB induced activation helps RPE to transitorily cope with the LB burden.

sensitize RPE cells to light exposure. Cellular photosensitivity is proportional to the amount of LBs accumulated [42] and the wavelength, with a maximum at 430 nm (blue light), which coincides with the excitation spectrum of the LBs [43]. The absorption of blue photons by the LBs' extended double bond conjugated system, in the presence of oxygen, leads to the formation of oxidized LB species [44, 45] that after repetitive oxidative attacks become fragmented into far reaching, highly reactive, carbonyl bearing small molecules [43, 44, 46–48]. These fragments promote cell damage by forming Schiff base adducts with free amine groups in lysosomal hydrolases, nucleotides, phospholipids, lipids, proteins [49], DNA [50], proteasomes [51], and molecules in extracellular retinal deposits (drusen), which could trigger local innate and adaptive immune responses [52, 53]. Interestingly, healthy mice immunized with Schiff base adducts found in the AMD lesions, developed AMD-like retinal pathology [54]. There is also *in vivo* evidence indicating that RPE lipofuscin undergoes photodegradation in the eye. Ueda et al [55] showed that $ABCA4^{-/-}$ animals were more susceptible to light damage than WT animals and that in both groups, older animals carrying larger amounts of LBs were also more susceptible. More recently Ref. [56] showed that RPE-lipofuscin photodegradation takes place in mouse eyes under standard ambient illumination. Specifically, they found that WT and $ABCA4^{-/-}$ mice reared in constant darkness contained 45 and 62% more LBs in the RPE than their respective 12-h cyclic light-reared controls. In addition, $ABCA4^{-/-}$ mice who received vitamin E, a potent inhibitor of LB oxidation [44], displayed 54% more LBs than controls. Studies in humans, using fluorescence microscopy for quantifying lipofuscin, and MALDI-IMS (high-resolution matrix assisted laser desorption-ionization imaging mass spectrometry) for detecting A2E showed that lipofuscin fluorescence colocalized with A2E only in the darkest zones of the retina [57]. When the same technology was applied to the eyes of $ABCA4^{-/-}$ mice, lipofuscin fluorescence, and A2E colocalized 100% [58]. Since RPE is exposed to higher levels of illumination in eyes from diurnal than nocturnal species, these data suggest that A2E is much more photooxidized into MALDI-IMS unidentifiable fluorescent derivatives in human eyes than in mice eyes. However, how much LB photooxidation contributes to retinal pathology is an open question. There is a large amount of clinical trial data on the use of antioxidants (lutein, zeaxanthin, and vitamins C and E) supplementation (alone or in combination) to prevent or delay retinal degeneration. A Cochrane meta-analysis performed on four large, high-quality–randomized clinical trials involving a total of 65,250 participants, without signs of AMD at baseline showed no effect of antioxidant therapy for preventing the onset of retinal degeneration *per se* [59]. Another Cochrane review meta-analysis [60] involving data from 13 randomized clinical trials, including two large trials, the AREDS1/2 and the Vitamin E Intervention in Cataract and Age-Related Maculopathy study, and 11 smaller (20–400 participants) randomized trials were performed to decide whether antioxidants can slow progression of retinal damage in patients with established AMD. The AREDS1/2 shows that long-term, high-dose supplementation with vitamin E (400 IU), vitamin C (500 mg), beta-carotene (15 mg), zinc (80 mg), and copper (2 mg) reduced the risk of progression to geographic atrophy AMD by 8% in only a subgroup of patients with intermediate AMD at baseline. The other 11 trials demonstrate little evidence for the effectiveness of antioxidant therapy for preventing either visual loss or AMD progression. In summary, treatments with antioxidants have shown very modest efficacy at preventing or stopping the progression of lypofuscin-associated retinal degenerations. Indeed, patients with mutations in ABCA4 (Stargardt-1, CRD, and RP) gene are not cured by high-doses of antioxidants [61]. This may indicate that scavenging reactive oxygen species is not the best approach to halt LB-driven damage. Alternatively, LB-photooxidation could be damaging through its propensity to activate the complement system [62]. In support of this idea, there is a histologic evidence of complement deposition in drusen of retinas with AMD [63] and animal studies show that overexpression of inhibitors of complement protects retinas of mice with elevated LB content [64]. Furthermore, genetic polymorphism in genes encoding complement factor H (CFH), CFB component C2, CFI, and complement components 2, 3 and 7 has been associated with elevated risk for LB-driven retina disease. Initially, there was a tremendous excitement to test complement inhibitors in the eye. Out of a dozen tested, only 1 molecule, Lampalizumab (Genentech), has made it into phase 3 clinical trials. In the phase 2 clinical test, Lampalizumab decreased the rate of growth of the geographic atrophy area, especially those with CFI polymorphisms [65]. Two world-wide multi-center prospective phase 3 clinical trials, which are enriched with CFI subjects, are now fully enrolled and results are expected in early 2018.

1.6.2. Inactivation of lysosome-dependent degradative processes

Because RPE cells are the most active postmitotic phagocytes in the body, they heavily rely on the fitness of their degradative machinery to operate. Indeed, a high baseline of autophagic activity level has been detected in the RPE and photoreceptors [66, 67], which were further enhanced during periods of POS phagocytosis [68]. Digestion of rhodopsin is also necessary for adaptation of rods to changes in light intensity [69]. Chemical or genetic inhibition of autophagy in RPE cells increased accumulation of undigested material and reduced cell viability [70]. Deletion of the autophagy inducer gene RB1CC1 in rodent RPE caused severe retinal degeneration, underlining the importance of basal autophagy [71]. Histological examination of retinas from Stargardt and AMD subjects revealed massive accumulation of lysosomal material similar to lipofuscin in the apical regions of RPE cells and of extruded extracellular deposits (drusen and pseudo-drusen) that support the idea of a defect in the recycling of endocytic and autophagic cargoes [72, 73]. *In vitro* experiments in which exogenous A2E was loaded in the lysosomes of cultured RPE cells, as surrogate of lipofuscin accumulation, show also a significant impairment in the digestion of phagocytized POS [74, 75] and autophagocytized proteins [76], implying lysosome-dependent degradative pathways are a primary point of attack by LB accumulation. The mechanism by which LBs mediate these inhibitions is not fully understood yet. Measurements of lysosomal protease, lipidase, glycosidase, nucleases, sulfatase, and phosphatase activities in homogenates of RPE revealed that A2E does not inhibit lysosomal activity by direct interaction with the hydrolases [77]. Lysosomes-containing A2E seems to have increased pH [74]. Bergmann et al. [76], working with purified lysosomes, provided evidence that A2E inhibits the vacuolar H(+)-ATPase (v-ATPase). v-ATPase is a transmembrane lysosomal protein in charge of maintaining the acidic environment within the lysosomes. Because acidic conditions are a prerequisite for the activity of lysosomal hydrolases, A2E-induced increase of lysosomal pH would explain, in part, its effect on lysosomal functions and autophagy [78]. Consistently, restoration of acidic pH in RPE lysosomes has shown promising results at improving lysosomal dependent degradative processes [79]. Furthermore, v-ATPase, together with mTORC1 complex, Rag GTPases, Ragulator, and Rheb, is an essential component of the lysosome nutrient-sensing (LYNUS) complex [80]. Under conditions of plenty of food, the v-ATPase complex senses luminal amino acids [81] and recruits mTORC1 to the lysosomal surface where it gets activated by phosphorylation [82]. Active mTORC1 complex is the main kinase negatively controlling autophagy and lysosomal biogenesis. When v-ATPase is inhibited by starvation, mTORC1 is released from the lysosome, becomes immediately inactive by dephosphorylation, and can no longer inhibit autophagy or TFEB nuclear translocation [83, 84]. The latter, by increasing lysosomal number, trafficking, hydrolase content, initiation of autophagy [85, 86], and lipid catabolism [87], facilitates the rapid degradation of a variety of substrates. How A2E inhibition of v-ATPase affects these cascades is not yet understood. Few studies have characterized the status of endogenous mTORC1 and TFEB in the RPE [88–90] and no enough data are available for LB-loaded RPE. A likely scenario is that TFEB activation by LBs provides a first line of defense that is insufficient to address accumulating autophagosomes containing partially degraded POS. However, in the absence of such lysosomal stress response, the RPE might succumb even faster. This model would explain experimental data showing that A2E induced a concentration- and time-dependent protective autophagic response in RPE cell cultures. [91]. Clinical trials using rapamycin, a mTORC1 specific inhibitor, to treat advanced stages of AMD showed no positive results [92, 93].

1.6.3. Lysosomal membrane permeabilization (LMP)

The A2E molecule contains a central pyridinium ring that houses permanently positive amine nitrogen and two long hydrophobic polyene arms. A similar structure is shared by other LBs, including A2-GPE, A2PE, and their isomers. Instead, A2-DHP-PE, all-trans-retinal (ATRD), all-trans-retinal dimer-E (ATRD-E), and all-trans-retinal dimer-PE (ATRD-PE) have non-charged ring cores, although ATRD-E and ATRD-PE have protonable nitrogens that confer them with amphipathic character at low pH. Amphiphatic LBs have the potential to intercalate into membranes [22, 94]. Schutt et al [95] investigated the destabilizing effects of A2E on purified lysosomes by measuring the release of luminal β-hexosaminidase to the supernatant. Concentrations as low as 2 μm induced leakage, whereas plasma membranes were insensitive to much higher concentrations. In support of the idea that amphipathic LBs cause lysosomal membrane permeabilization (LMP) is the observation that RPE cells loaded with lipofuscin granules or A2E into their lysosomes undergo significant LMP [40]. Multiple mechanisms can be responsible for A2E mediating LMP. A2E can act as a surfactant and cause direct membrane damage. De and Sakmar [94] found that A2E-induced leakage of liposomes at concentrations of 200–300 μm. LMP could also be the result of A2E crystallization within lysosomes, which might cause inflammation by activating a multimolecular signaling complex of the innate immune system, the NLRP3 inflammasome, resulting in a caspase-1–mediated activation and secretion of mature IL1β family cytokines [96, 97]. Relevantly, A2E accumulation induces NALP3-mediated secretion of mature IL1β [98]. Fluorescence staining of lipofuscin revealed a membrane bound autofluorescent granule with the bulk of A2E in the lumen rather than in the membrane. Atomic force microscopy shows the core of the granule comprises of solid mini aggregates [99]. Accordingly, we observed A2E (MW 592 Da) in aqueous media cannot cross 0.10 micron pore size filters with molecular weight cutoff of 300,000 Da. This retention was due to size exclusion as A2E passed through 3 micron filters of the same material (**Figure 5**).

Toxicity of lipofuscin could also involve other less-studied mechanisms, including ***mitochondrial poisoning***, as it has been shown that lysosomal A2E progressively leaks into the mitochondrial compartment [100], where it destabilizes the membrane [101] and inhibits oxidative phosphorylation [102], derail of **cholesterol trafficking** [103], ***activation of Retinoid Acid Receptor*** (***RAR***)-dependent VEGF secretion in RPE [104, 105], and ***inhibition of RPE-65*** isomerohydrolase activity, which limits the RPE supply of 11-cis retinal [106]. In summary, although numerous mechanisms of toxicity elicited by pathologic accumulation of LBs in RPE lysosomes have been proposed, no viable therapeutic options have resulted yet from targeting them. Hence, strategies to reduce LB accumulation from RPE have been further investigated.

2. Strategies to reduce lipofuscin accumulation in RPE cells

Alternative strategies to mitigate the cytotoxic effects of LBs involve preventing their accumulation. Two approaches have been pursued (1) to prevent *de novo* formation of LB and (2) to remove previously accumulated LB.

2.1. Strategies that prevent *de novo* formation of LBs

Long-term restriction of vitamin A intake has been shown to reduce retinaldehyde levels in RPE but is not a therapeutic option, since it causes night blindness and systemic hypovitaminosis [107]. In 2005, Radu et al. [108] showed that oral administration of a synthetic form of vitamin A (***fenretinide***), already in use against cancer, acne, cystic fibrosis, rheumatoid arthritis, and psoriasis, could **competitively block RBP4 transport of vitamin A from the**

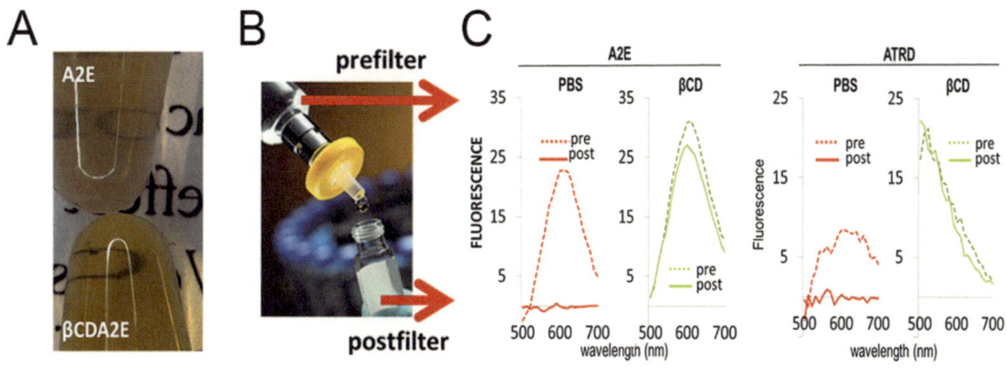

Figure 5. LBs form crystal-like aggregates in aqueous environment. (A) Turbidity of A2E dilutions in water versus 0.1 M βCD. (B) Passage through 0.10 micron pore filters (MW cutoff of 300,000 Da) of LBs in water or βCD solutions. (C) Autofluorescent detection of A2E (MW 592 Da) and ATRD (600 Da) in pre and post filtrates. Retention was due to size exclusion since they crossed 3 micron filters of identical material (not shown).

blood to the RPE. Oral fenretinide produced mild reversible skin dryness and night blindness. However, in 2011, a phase-2 study on 225 AMD patients failed to show beneficial effects. **Oral emixustat hydrochloride** is a synthetic nonretinoid reversible inhibitor of the RPE65 enzyme, which converts all-trans-retinyl to 11-cis-retinal, a rate-limiting reaction of the visual cycle. This drug showed minimal toxicity in phase-1 trials and effectively reduced photoreceptor response to light, consistent with its mechanism of action. However, in May 2016, the results from the phase 2b/3 SEATTLE study did not show any significant difference in retinal degenerative rate or visual acuity changes. **Oral deuterated vitamin A** (ALK-001), is vitamin A modified by replacing hydrogen with deuterium, a safe, nonradioactive isotope. Deuterated vitamin A has lower tendency to spontaneously dimerize into LBs. Long-term, oral administration of ALK-001 to $ABCA4^{-/-}$ reduced the accumulation of lipofuscin and A2E by 70 and 80%, respectively [109]. Assessment of the retina electric response to light signals (electroretinogram) revealed that ALK-001 treatment prevented the gradual loss of visual function observed in the $ABCA4^{-/-}$ mouse. Safety phase-1 clinical trials have been completed but phase-2 is ongoing. It is too early to know whether ALK-001 will be beneficial for Stargardt or AMD patients.

Oral aldehyde traps (VM200, Vision Medicines) constitute of new drugs that react with retinaldehydes forming reversible Schiff bases and thus, reducing the available levels of free aldehydes with cellular amine groups. In preclinical studies, VM200 preserved retinal structure and function of mice retinas in a dose-dependent manner [7]. Safety phase-1 is in progress but there is no effectiveness data in humans.

In summary, none of the visual cycle modulators have made it out of the nearly half a dozen phase 1 and phase 2 clinical trials, so far. They all cause significant night blindness that limit their use. Although they all seem very effective at slowing down the formation of new LBs in animals, they lack effect on previously accumulated LB, which may explain why fenretinide or emixustat did not benefit patients already diagnosed with AMD. Likewise in animal models, where these drugs are given preventively for long periods, humans may need to take them early, i.e., much before the clinical manifestations appear, and for life.

2.2. Strategies for removing previously accumulated LB

Potentially, their main advantage over agents that prevent accumulation of LBs is that they might be administered to patients with Stargardt or AMD who display large buildup of LBs. Most of these strategies are in preclinical stage.

2.2.1. Oral soraprazan (Katairo GmbH)

In 2012, Schraermeyer's group reported successful elimination of lipofuscin from RPE cells in monkey retinas after 1 year of oral administration of the drug. Researchers showed that lipofuscin granules were expelled by RPE cells toward their basolateral side and were cleared by macrophages recruited to the area [110]. Although soraprazan is known to reversibly block the potassium binding site of the gastric H+/K+ ATPase proton pump, the precise mechanism by which it causes clearance of LB deposits is unknown.

2.2.2. *Enzymatic degradation of LBs*

Because LBs are refractory to degradation by lysosomal hydrolases, several groups searched for exogenous enzymes with LB destroying activity. Horseradish peroxidase (HRP) was the first one identified [111]. It catalyzes the oxidative cleavage of the polyene–arms of LBs. To test the effectiveness of HRP, cultures of RPE cells preloaded with LBs underwent an enzyme replacement therapy-like treatment with HRP. The efficiency of the clearance was low but the major problem was the considerable amount of highly toxic reactive molecules released as a by-product of the HRP-mediated oxidation of LBs [112]. It was more recently reported in Ref. [113] that neutrophil myeloperoxidase (MPO) catalyzes the *in vitro* degradation of A2E. The authors delivered MPO to lysosomes of RPE cells via mannose-6-phosphate (M6P) receptor. M6P-MOP exhibited a half-life of 10 h in the lysosomes and degraded lysosomal A2E in, but also disrupted lysosomal acidification and triggered lysosomal stress, manifested by the nuclear translocation of TFEB that eventually led to cell death. Thus, the strategy of eliminating LBs with peroxidases seems to be limited by the inherent associated release of detrimental reactive species, which would be equivalent to try to clear LBs by photooxidation.

2.2.3. *Beta-cyclodextrins (β-CDs)*

βCDs are membrane-impermeant cyclic sugars made of seven glucose residues. They contain a nonpolar central cavity that is capable of accommodating hydrophobic ligands and a hydrophilic outer surface that makes them soluble in water [114]. Several FDA-approved cyclodextrins are currently used to improve the delivery of lipophilic drugs. We demonstrated that βCDs form soluble complexes with LBs [115] (**Figure 5**). *In silico* modeling predicted 2:1 βCD-A2E complex, where one βCD accommodates per arm of LB. We also observed that βCDs reduced the content of A2E from polarized RPE monolayers on Transwell filter cultures and from RPE in the eyes of mice that accumulate massive amounts of lysosomal LBs [116] indicating that βCD treatment can eliminate not only A2E, but also the complex LB mixes found in RPE lipofuscin. The mechanisms by which βCDs induce clearance of lysosomes' content in RPE is yet to be determined but could potentially be optimized to develop a novel therapeutic approach to clear LB-buildups. The mechanism of cholesterol removal by βCDs is one of the best characterized. Likewise LBs, cholesterol forms soluble inclusion complexes with βCDs. In normal cells, cholesterol is more abundantly present in plasma membrane, common recycling endosome and trans-Golgi complex [117]. Removal of cholesterol from these membranes, requires high concentrations (5–10 mM) and prolonged times because βCDs, which have an ~8 Å deep cavity, must form stacked dimers (improbable event) to remove 18 Å long cholesterols out of the lipid bilayer and shelter them from the water [118]. In cells with Niemann Pick Type-C defect, i.e., with inactivating mutations in either NPC1 or NPC2 genes, that code for two intra-lysosomal lipid transporter proteins, cholesterol is, also, aberrantly accumulated within lysosomes. Removal of lysosomal cholesterol buildups requires lower concentrations (0.1–1.0 mM) and shorter incubation times with βCD [119]. Furthermore, sulfo-butyl-ether-βCD (Captisol®), a βCD derivative that cannot form stacked dimers and that therefore cannot solubilize membrane cholesterol, can still reduce lysosomal buildups [120]. The model for βCD-mediated removal of lysosomal cholesterol proposes that βCD enters lysosomes by endocytosis, where it binds free cholesterol in the lumen and shuttle it to the limiting lysosomal membrane [121, 122]. From there, cholesterol is transferred, by a not fully characterized trafficking machinery, that probably involves points of membrane contact between organelles and cholesterol binding proteins [123], to the ER, plasma membrane, peroxisomes, and mitochondria [124, 125]. In the absence of extracellular cholesterol acceptor molecules, the stoichiometric analysis of βCD clearance provides no evidence of cholesterol release to the media but rather indicates a rapid metabolic processing within the cytosolic compartment [126, 127]. In the case of LBs, if βCDs clearance works similarly, then it would be important to see what putative LB-transport system acts thereafter to ship LBs for degradation [85, 128] or to expel them from the cell. Confirmation of such operating trafficking pathway could represent an important advance to identify pharmacological targets for the elimination of lysosomal LBs.

In a mice model of atherosclerosis [129], βCDs have shown effective removal of cholesterol crystals from macrophage foam cells. The mechanism in this case seems to be mostly mediated by the execution of a LXR (liver-X-receptor) dependent transcriptional program response that enhanced the efflux and degradation of cholesterol and reduced inflammation.

Administration of βCD also lowered the levels of amyloid-β in an animal model of Alzheimer Disease [130] and from drusen deposits in animal models of Stargardt disease [131]. The mechanism in these cases is less clear but it seems to be transcriptionally controlled. Accordingly, it will be important to determine if βCDs trigger a transcriptional program that primes RPE cells to eliminate its lysosomal content, independently on whether they form soluble complex with the wasted material. Similarly, fibroblasts from patients with ceroid-lipofuscinosis, the most common cause of neurodegeneration of children in the United States, and cellular or animal models with misfolded α-synuclein accumulation were cleared by βCDs [132, 133]. The mechanism in these cases seemed to be mediated by TFEB [86, 134]. The pathway responsible for βCD activation of TFEB is not defined and is induced by millimolar doses of cyclodextrins. βCDs appear to induce autophagy [135] and exocytosis of lysosomes [136]. Finally, overexpression of activated TFEB has been demonstrated to ameliorate pathology in late-onset neurodegenerative diseases such as Parkinson, Huntington and Alzheimer, as well as in models of spinal and bulbar muscular atrophy and to clear deposits in lysosomal storage disorders (LSDs) [83, 86, 137–139]. Thus, it will be important to determine whether TFEB is necessary or if it can synergistically contribute to the clearance of RPE lipofuscin.

3. Conclusions

Buildup of lipofuscin in RPE lysosomes often evolves into irreversible damage of overlying photoreceptors. This is a common event in individuals with mutations in ABCA4 gene and is believed to underlie the progression of age-related lesions in AMD people, the most common cause of blindness in the elderly population. Unfortunately, the great majority of patients diagnosed with this problem have no therapeutic options available. Analysis of RPE lipofuscin, identified as major components lipid bisretinoids (LBs), sub-products of the spontaneous dimerization of retinaldehydes produced during the visual cycle. Targeting LBs secondary complications with oral antioxidants, inhibitors of complement or autophagy inducers provided no or little beneficial effect. The recent failures in clinical trials with visual cycle modulators, which prevent *de novo* formation of LBs may reflect the incapacity of these drugs to stop degeneration once LB-accumulation has been established, which is probably the case for most individuals with clinical symptoms. Hence, the development of novel strategies to permanently remove previously accumulated lipofuscin is an urgent medical need. The characterization of LBs as the core-components of RPE lipofuscin has permitted to rationally develop strategies to remove them from RPE cells in the laboratory. This, combined with an improved understanding of the molecular pathways that govern autophagy and stimulate cellular clearance might allow in the near future, to develop improved therapies for retinal degenerations resulting from genetic or age-related retinal lipofuscin accumulation.

Acknowledgements

This study was supported by grant R01 EY027422 from NIH and BrightFocus Foundation award M2016124 to M.M.N, by grant RO1 EY08538 to ERB and by Departmental grants from Research to Prevent Blindness Foundation and Dyson Foundation.

Author details

Marcelo M. Nociari*, Szilard Kiss and Enrique Rodriguez-Boulan*

*Address all correspondence to: mnociari@med.cornell.edu and boulan@med.cornell.edu

Department of Ophthalmology, Weill Cornell Medicine, Margaret Dyson Vision Research Institute, Cornell University, New York, NY, USA

References

[1] Bok D. The retinal pigment epithelium: a versatile partner in vision, J. Cell Sci. 1993;189 LP-195

[2] Streilein JW, Ma N, Wenkel H, Fong Ng T, Zamiri P. Immunobiology and privilege of neuronal retina and pigment epithelium transplants. Vision Research. 2002;**42**:487-495

[3] Wald G. The molecular basis of visual excitation. Nature. 1968;**219**:800-807

[4] Fu Y. Phototransduction in rods and cones. In: Webvision: The Organisation of the Retina and Visual System. 2010: pp. 1-47

[5] Palczewski K. JBC Papers in Press. Published on November 10, 2011 as Manuscript R111. 301150 The latest version is at http://www.jbc.org/cgi/doi/10.1074/jbc.R111.301150. 2011. DOI: 10.1074/jbc.R111.301150

[6] Wang JS, Kefalov VJ. The cone-specific visual cycle. Progress in Retinal and Eye Research. 2011;**30**:115-128. DOI: http://dx.doi.org/10.1016/j.preteyeres.2010.11.001

[7] Maeda A, Golczak M, Chen Y, Okano K, Kohno H, Shiose S, Ishikawa K, Harte W, Palczewska G, Maeda T, Palczewski K. Primary amines protect against retinal degeneration in mouse models of retinopathies. Nature Chemical Biology. 2012;**8**:170-178. DOI: 10.1038/nchembio.759

[8] Boyer NP, Higbee D, Currin MB, Blakeley LR, Chen C, Ablonczy Z, Crouch RK, Koutalos Y. Lipofuscin and N-retinylidene-N-retinylethanolamine (A2E) accumulate in retinal pigment epithelium in absence of light exposure: Their origin is 11-cis-retinal. Journal of Biological Chemistry. 2012;**287**:22276-22286. DOI: 10.1074/jbc.M111.329235

[9] Wu Y, Li J, Yao K. Structures and biogenetic analysis of lipofuscin bis-retinoids. Biomedical & Biotechnology. 2013;**14**:763-773. DOI: 10.1631/jzus.B1300051

[10] Young RW. The renewal of photoreceptor cell outer segments. Journal of Cell Biology. 1967;**33**:61-72

[11] McMahon DG, Iuvone PM, Tosini G. Circadian organization of the mammalian retina: From gene regulation to physiology and diseases. Progress in Retinal and Eye Research. 2014;**0**:58-76. DOI: 10.1016/j.preteyeres.2013.12.001

[12] Kim J, Zhao H, Martinez J, Doggett TA, Alexander V, Tang PH, Ablonczy Z, Chan CC, Zhou Z, Green R, Ferguson TA. Non-canonical autophagy promotes the visual cycle. Cell. 2014;**154**:365-376. DOI: 10.1016/j.cell.2013.06.012.Non-canonical

[13] Mao Y, Finnemann SC. Regulation of phagocytosis by Rho GTPases. Small GTPases. 2015;**6**:1-11. DOI: 10.4161/21541248.2014.989785

[14] Croce AC, Bottiroli G. Autofluorescence spectroscopy and imaging: A tool for biomedical research and diagnosis. European Journal of Histochemistry. 2014;**58**:2461. DOI: 10.4081/ejh.2014.2461

[15] Farin S, Porta EA. Sequential histochemical studies of neuronal lipofuscin in human cerebral cortex from the first to the ninth decade of life. Archives of Gerontology and Geriatrics. 2002;**34**:219-231

[16] Brunk UT, Terman A. Lipofuscin: Mechanisms of age-related accumulation and influence on cell function. Free Radical Biology and Medicine. 2002;**33**:611-619. DOI: 10.1016/S0891-5849(02)00959-0

[17] Delori FC, Dorey CK, Staurenghi G, Arend O, Goger DG, Writer JJ. In vivo fluorescence of the ocular fundus exhibits retinal pigment epithelium lipofuscin characteristics. Investigative Ophthalmology & Visual Science. 1995;**36**:718-729

[18] Weirer JJ, Delori FC, Wing GL, Fitch KA. Retinal pigment epithelial lipofuscin and melanin and choroidal melanin in human eyes. Investigative Ophthalmology & Visual Science. 1986;**27**:145-152

[19] Parish CA, Hashimoto M, Nakanishi K, Dillon J, Sparrow J. Isolation and one-step preparation of A2E and iso-A2E, fluorophores from human retinal pigment epithelium. Proceedings of the National Academy of Sciences of United States of America. 1998;**95**:14609-14613

[20] Feeney-Burns L, Hilderbrand ES, Eldridge S. Aging human RPE: Morphometric analysis of macular, equatorial, and peripheral cells. Investigative Ophthalmology & Visual Science. 1984;**25**:195-200

[21] Eldred GE, Katz MI. The autofluorescent products of lipid peroxidation may not be lipofuscin-like. Free Radical Biology & Medicine. 1989;**7**:157-163. DOI: 10.1016/0891-5849(89)90007-5

[22] Eldred GE, Lasky MR. Retinal age pigments generated by self-assembling lysosomotropic detergents. Nature. 1993;**361**:724-726

[23] Ren RX, Sakai N, Nakanishi K. Total synthesis of the ocular age pigment A2-E: A convergent pathway. Journal of the American Chemical Society. 1997;**7863**:3619-3620

[24] Sparrow JR, Kim SR, Wu Y. Experimental approaches to the study of A2E, a bisretinoid lipofuscin chromophore of retinal pigment epithelium. Methods in Molecular Biology. 2010;**652**:315-327. DOI: 10.1007/978-1-60327-325-1_18

[25] Boulton ME. Studying melanin and lipofuscin in RPE cell culture models. Experimental Eye Research. 2014;**126**:61-67. DOI: 10.1016/j.exer.2014.01.016

[26] Kiser PD, Golczak M, Palczewski K. Chemistry of the retinoid (visual) cycle. Chemical Reviews. 2013;**114**:194-232. DOI: 10.1021/cr400107q

[27] Kim SR, Jang YP, Jockusch S, Fishkin NE, Turro NJ, Sparrow JR. The all-trans-retinal dimer series of lipofuscin pigments in retinal pigment epithelial cells in a recessive Stargardt disease model. Proceedings of the National Academy of Sciences of United States of America. 2007;**104**:19273-19278. DOI: 10.1073/pnas.0708714104

[28] Sparrow JR, Gregory-Roberts E, Yamamoto K, Blonska A, Ghosh SK, Ueda K, Zhou J. The bisretinoids of retinal pigment epithelium. Progress in Retinal and Eye Research. 2012;**31**:121-135. DOI: 10.1016/j.preteyeres.2011.12.001

[29] Murdaugh LS, Dill AE, Dillon J, Simon JD, Gaillard ER. Age-related changes in rpe lipofuscin lead to hydrophobic polymers. In: Studies on Retinal and Choroidal Disorders. 2012:pp. 113-139. DOI: 10.1007/978-1-61779-606-7

[30] Schutt F, Ueberle B, Schno M, Holz FG, Kopitz J. Proteome analysis of lipofuscin in human retinal pigment epithelial cells. FEBS Letters. 2002;**528**:217-221

[31] Schutt F, Bergmann M, Holz FG, Kopitz J. Proteins modified by malondialdehyde, 4-hydroxynonenal , or advanced glycation end products in lipofuscin of human retinal pigment epithelium. Investigative Ophthalmology & Visual Science. 2017;**44**:3663-3668. DOI: 10.1167/iovs.03-0172

[32] Warburton S, Southwick K, Hardman RM, Secrest AM, Grow RK, Xin H, Woolley AT, Burton GF, Thulin CD. Examining the proteins of functional retinal lipofuscin using proteomic analysis as a guide for understanding its origin. Molecular Vision. 2005;**11**: 1122-1134

[33] Brunet S, Thibault P, Gagnon E, Kearney P, Bergeron JJM, Desjardins M. Organelle proteomics : Looking at less to see more. Trends in Cell Biology. 2003;**13**:629-638. DOI: 10.1016/j.tcb.2003.10.006

[34] Ng K, Gugiu B, Renganathan K, Davies MW, Gu X, Crabb JSWSW, Kim SR, Rózanowska MB, Bonilha VL, Rayborn ME, Salomon RG, Sparrow JR, Boulton ME, Hollyfield JG, Crabb JSWSW, Malgorzata B. Retinal pigment epithelium lipofuscin proteomics. Molecular & Cellular Proteomics. 2008;**7**:1397-1405. DOI: 10.1074/mcp.M700525-MCP200

[35] H.E. Bazan, N.G. Bazan, L. Feeney-Burns, E.R. Berman, Lipids in human lipofuscin-enriched subcellular fractions of two age populations. Comparison with rod outer segments and neural retina., Invest. Ophthalmol. Vis. Sci. 1990;**31**:1433-1443

[36] Katz ML, Redmond TM. Effect of Rpe65 knockout on accumulation of lipofuscin fluorophores in the retinal pigment epithelium. Investigative Ophthalmology & Visual Science. 2001;**42**

[37] Molday RS, Zhong M, Quazi F. The role of the photoreceptor ABC transporter ABCA4 in lipid transport and Stargardt macular degeneration. Biochimica et Biophysica Acta. 2009;**1791**:573-583. DOI: 10.1016/j.bbalip.2009.02.004

[38] Mihai DM, Washington I. Vitamin A dimers trigger the protracted death of retinal pigment epithelium cells. Cell Death & Disease. 2014;**5**:e1348. DOI: 10.1038/cddis.2014.314

[39] Rózanowska M, Jarvis-Evans J, Korytowski W, Boulton M, Burke J, Sarna T. Blue light-induced reactivity of retinal age pigment. Journal of Biological Chemistry. 1995:18825-18830

[40] Schu F, Davies S, Holz FG, Boulton ME. Photodamage to human RPE cells by A2-E, a retinoid component of lipofuscin. Investigative Ophthalmology & Visual Science. 2000;**41**:2303-2308

[41] Yamamoto K, Yoon KD, Ueda K, Hashimoto M, Sparrow JR. A novel bisretinoid of retina is an adduct on glycerophosphoethanolamine. Investigative Ophthalmology & Visual Science. 2011;**52**:9084-9090. DOI: 10.1167/iovs.11-8632

[42] Sparrow JR, Nakanishi K, Parish CA. The lipofuscin fluorophore A2E mediates blue light-induced damage to retinal pigmented epithelial cells. Investigative Ophthalmology & Visual Science. 2000;**41**:1981-1989

[43] Boulton M, Dontsov A, Jarvis-Evans J, Ostrovsky M, Svistunenko D. Lipofuscin is a photoinducible free radical generator. Journal of Photochemistry & Photobiology B. 1993;**19**:201-204

[44] Sparrow JR, Zhou J, Ben-Shabat S, Vollmer H, Itagaki Y, Nakanishi K. Involvement of oxidative mechanisms in blue-light-induced damage to A2E-laden RPE. Investigative Ophthalmology & Visual Science. 2002;**43**:1222-1227

[45] Ben-shabat S, Itagaki Y, Jockusch S, Sparrow JR, Turro NJ, Nakanishi K. Formation of a nonaoxirane from A2E, degeneration, and evidence of singlet oxygen involvement. Angewandte Chemie (International Ed. in English). 2002;**41**:814-817

[46] Washington I, Jockusch S, Itagaki Y, Turro NJ, Nakanishi K. Superoxidation of bis-retinoids. Angewandte Chemie (International Ed. in English). 2005;**44**:7097-7100. DOI: 10.1002/anie.200501346

[47] Yoon KD, Yamamoto K, Ueda K, Zhou J, Sparrow JR. A novel source of methylglyoxal and glyoxal in retina: Implications for age-related macular degeneration. PLoS One. 2012;**7**:e41309. DOI: 10.1371/journal.pone.0041309

[48] Wang Z, Keller LMM, Dillon J, Gaillard ER, Oxidation of A2E results in the formation of highly reactive aldehydes and ketones. Photochemistry and Photobiology. 2006;82:1251-1257. DOI: 10.1562/2006-04-01-RA-864

[49] Yoon KD, K Yamamoto, J Zhou, JR Sparrow, Photo-products of retinal pigment epithelial bisretinoids react with cellular thiols, 2011 1839-1849

[50] Sparrow JR, Zhou J, Cai B. DNA is a target of the photodynamic effects elicited in A2E-laden RPE by blue-light illumination. Investigative Ophthalmology & Visual Science. 2003;**44**:2245. DOI: 10.1167/iovs.02-0746

[51] Zhang X, Zhou J, Fernandes AF, Sparrow JR, Pereira P, Taylor A, Shang F. The proteasome: A target of oxidative damage in cultured human retina pigment epithelial cells. Investigative Ophthalmology & Visual Science. 2008;**49**:3622-3630. DOI: 10.1167/iovs.07-1559

[52] J. Zhou, K. Ueda, J. Zhao, J.R. Sparrow, Correlations between Photodegradation of Bisretinoid Constituents of Retina and Dicarbonyl Adduct Deposition *, J. Biol. Chem. 2015;**290**:27215-27227

[53] Kanda A, Abecasis G, Swaroop A. Inflammation in the pathogenesis of age-related macular degeneration. British Journal of Ophthalmology. 2008;**92**:448-450. DOI: 10.1136/bjo.2007.131581

[54] Hollyfield JG, Bonilha VL, Rayborn ME, Yang X, Shadrach KG, Lu L, Ufret RL, Salomon RG, Perez VL. Oxidative damage-induced inflammation initiates age-related macular degeneration. Nature Medicine. 2008;**14**:194-198. DOI: 10.1038/nm1709

[55] Wu L, Ueda K, Nagasaki T, Sparrow JR, Light damage in Abca4 and Rpe65rd12 mice. Investigative Ophthalmology & Visual Science. 2014;**55**:1910-1918. DOI: 10.1167/iovs.14-13867

[56] K. Ueda, J. Zhao, H.J. Kim, J.R. Sparrow, H. Jin, J.R. Sparrow, Photodegradation of retinal bisretinoids in mouse models and implications for macular degeneration., Proc. Natl. Acad. Sci. U. S. A. 2016;**113**:6904-6909

[57] Ablonczy Z, Higbee D, Anderson DM, Dahrouj M, Grey AC, Gutierrez D, Koutalos Y, Schey KL, Hanneken A, Crouch RK. Lack of correlation between the spatial distribution of

A2E and lipofuscin fluorescence in the human retinal pigment epithelium. Investigative Ophthalmology & Visual Science. 2013;**54**:5535-5542. DOI: 10.1167/iovs.13-12250

[58] Grey AC, Crouch RK, Koutalos Y, Schey KL, Ablonczy Z. Spatial localization of A2E in the retinal pigment epithelium. Investigative Ophthalmology & Visual Science. 2011;**52**:3926-3933. DOI: 10.1167/iovs.10-7020

[59] J.R. Evans, J.G. Lawrenson, K.S. Henshaw, J.G. Lawrenson, Antioxidant vitamin and mineral supplements for preventing age-related macular degeneration, Cochrane Libr. 2012;**11**:CD000254

[60] Evans JR, Lawrenson JG. Antioxidant vitamin and mineral supplements for slowing the progression of age-related macular degeneration. Cochrane Database Systems Reviews. 2012;**11**:CD000254. DOI: 10.1002/14651858.CD000254.pub3

[61] Downie LE, Keller PR. Degeneration : Research evidence in practice. Optometry and Vision Science. 2014;**91**:821-831

[62] Zhou J, Kim SR, Westlund BS, Sparrow JR. Complement activation by bisretinoid constituents of RPE lipofuscin. Investigative Ophthalmology & Visual Science. 2009;**50**:1392-1399. DOI: 10.1167/iovs.08-2868

[63] Bradley DT, Zipfel PF, Hughes AE. Complement in age-related macular degeneration : A focus on function. Eye. 2011;**25**:683-693. DOI: 10.1038/eye.2011.37

[64] T.L. Lenis, S. Sarfare, Z. Jiang, M.B. Lloyd, D. Bok, R.A. Radu, Complement modulation in the retinal pigment epithelium rescues photoreceptor degeneration in a mouse model of Stargardt disease., Proc. Natl. Acad. Sci. U. S. A. 2017;**114**:3987-3992

[65] Hanus J, Zhao F, Wang S. Current therapeutic developments in atrophic age-related macular degeneration. 2016:122-127. DOI: 10.1136/bjophthalmol-2015-306972

[66] Chen Y, Sawada O, Kohno H, Le YZ, Subauste C, Maeda T, Maeda A. Autophagy protects the retina from light-induced degeneration. Journal of Biological Chemistry. 2013;**288**:7506-7518. DOI: 10.1074/jbc.M112.439935

[67] Mitter SK, Rao HV, Qi X, Cai J, Sugrue A, Dunn WA, Grant MB, Boulton ME. Autophagy in the retina: A potential role in age-related macular degeneration, In: LaVail MM, Ash JD, Anderson RE, Hollyfield JG, Grimm C (Eds.), Retinal Degenerative Diseases. Springer US: Boston, MA;2012: pp. 83-90. DOI: 10.1007/978-1-4614-0631-0_12

[68] Yao J, Jia L, Shelby SJ, Ganios AM, Feathers K, Thompson DA, Zacks DN. Circadian and noncircadian modulation of autophagy in photoreceptors and retinal pigment epitheliumcircadian and noncircadian modulation of autophagy. Investigative Ophthalmology & Visual Science. 2014;**55**:3237-3246

[69] Remé CE, Wolfrum U, Imsand C, Hafezi F, Williams TP. Photoreceptor autophagy: Effects of light history on number and opsin content of degradative vacuoles. Investigative Ophthalmology & Visual Science. 1999;**40**:2398-2404

[70] Mitter SK, Song C, Qi X, Mao H, Rao H, Akin D, Lewin A, Grant M, Dunn W, Ding J, Bowes Rickman C, Boulton M. Dysregulated autophagy in the RPE is associated with increased susceptibility to oxidative stress and AMD. Autophagy. 2014;**10**:1989-2005. DOI: 10.4161/auto.36184

[71] Yao J, Jia L, Khan N, Lin C, Mitter SK, Boulton ME, Dunaief JL, Klionsky DJ, Guan JL, Thompson DA, Zacks DN. Deletion of autophagy inducer RB1CC1 results in degeneration of the retinal pigment epithelium. Autophagy. 2015;**11**:939-953. DOI: 10.1080/15548627.2015.1041699

[72] Eagle RC Jr., Lucier AC, Bernardino VB Jr., Yanoff M. Retinal pigment epithelial abnormalities in fundus flavimaculatus: A light and electron microscopic study. Ophthalmology. 1980;**87**:1189-1200

[73] T. Ach, E. Tolstik, J.D. Messinger, A. V Zarubina, R. Heintzmann, C.A. Curcio, Lipofuscin Redistribution and Loss Accompanied by Cytoskeletal Stress in Retinal Pigment Epithelium of Eyes With Age-Related Macular Degeneration, Invest. Ophthalmol. Vis. Sci. 2015;**56**:3242-3252

[74] Holz FG, Schütt F, Kopitz J, Eldred GE, Kruse FE, Völcker HE, Cantz M. Inhibition of lysosomal degradative functions in RPE cells by a retinoid component of lipofuscin. Investigative Ophthalmology & Visual Science. 1999;**40**:737-743

[75] Vives-bauza C, Anand M, Shirazi AK, Magrane J, Gao J, Vollmer-snarr HR, Manfredi G, Finnemann SC, Shiraz AK, Shirazi AK, Magrane J, Gao J, Vollmer-snarr HR, Manfredi G, Finnemann SC. The age lipid A2E and mitochondrial dysfunction synergistically impair phagocytosis by retinal pigment epithelial cells. Journal of Biological Chemistry. 2008;**283**:24770-24780. DOI: 10.1074/jbc.M800706200

[76] M. Bergmann, F. Schütt, F.G. Holz, J. Kopitz, Inhibition of the ATP-driven proton pump in RPE lysosomes by the major lipofuscin fluorophore A2-E may contribute to the pathogenesis of age-related macular degeneration, FASEB J. 2004;**8**:562-564

[77] Bermann M, Schütt F, Holz FG, Kopitz J. Does A2E, a retinoid component of lipofuscin and inhibitor of lysosomal degradative functions, directly affect the activity of lysosomal hydrolases?. Experimental Eye Research. 2001;**72**:191-195. DOI: 10.1006/exer.2000.0949

[78] Yamamoto A, Tagawa Y, Yoshimori T, Moriyama Y, Masaki R, Tashiro Y. Bafilomycin A1 prevents maturation of autophagic vacuoles by inhibiting fusion between autophagosomes and lysosomes in rat hepatoma cell line, H-4-II-E cells. Cell Structure and Function. 1998;**23**:33-42. DOI: 10.1247/csf.23.33

[79] Guha S, Liu J, Baltazar G, Laties AM, Mitchell CH. Rescue of compromised lysosomes enhances degradation of photoreceptor outer segments and reduces lipofuscin-like autofluorescence in retinal pigmented epithelial cells. Advances in Experimental Medicine and Biology. 2014;**801**:105-111. DOI: 10.1007/978-1-4614-3209-8_14

[80] Zoncu R, Bar-Peled L, Efeyan A, Wang S, Sancak Y, Sabatini DM. mTORC1 senses lysosomal amino acids through an inside-out mechanism that requires the vacuolar H+-ATPase. Science (80-.). 2011;**334**:678-683. DOI: 10.1126/science.1207056

[81] Sancak Y, Peterson TR, Shaul YD, Lindquist RA, Thoreen CC, Bar-Peled L, Sabatini DM. The Rag GTPases bind raptor and mediate amino acid signaling to mTORC1. Science (80-.). 2008;**320**:1496-1501

[82] Benjamin D, Hall MN. mTORC1: turning off is just as important as turning on. Cell. 2014;**156**:627-628. DOI: http://dx.DOI.org/10.1016/j.cell.2014.01.057

[83] Sardiello M, Palmieri M, di Ronza A, Medina DL, Valenza M, Gennarino VA, Di Malta C, Donaudy F, Embrione V, Polishchuk RS, Banfi S, Parenti G, Cattaneo E, Ballabio A. A gene network regulating lysosoal biogenesis and function. Science (80-.). 2009;**325**:473-478

[84] Palmieri M, Impey S, Kang H, di Ronza A, Pelz C, Sardiello M, Ballabio A,.Characterization of the CLEAR network reveals an integrated control of cellular clearance pathways. Human Molecular Genetics. 2011;**20**:3852-3866. DOI: 10.1093/hmg/ddr306

[85] Settembre C, Fraldi A, Medina DL, Ballabio A, Signals from the lysosome: A control centre for cellular clearance and energy metabolism. Nature Reviews Molecular Cell Biology. 2013;**14**:283-296. DOI: 10.1038/nrm3565

[86] Medina DL, Fraldi A, Bouche V, Annunziata F, Mansueto G, Spampanato C, Puri C, Pignata A, Martina JA, Sardiello M, Palmieri M, Polishchuk R, Puertollano R, Ballabio A. Transcriptional activation of lysosomal exocytosis promotes cellular clearance. Developmental Cell. 2011;**21**:421-430. DOI: 10.1016/j.devcel.2011.07.016

[87] Ghosh A, Pahan K. PPARα in lysosomal biogenesis: A perspective. Pharmacological Research. 2016;**103**:144-148. DOI: 10.1016/j.phrs.2015.11.011

[88] Yu B, Xu P, Zhao Z, Cai J, Sternberg P, Chen Y. Subcellular distribution and activity of mechanistic target of rapamycin in aged retinal pigment epithelium dysregulated mTORC1 signaling in aged RPE. Investigative Ophthalmology & Visual Science. 2014;**55**:8638-8650

[89] Chen Y, Wang J, Cai J, Sternberg P. Altered mTOR signaling in senescent retinal pigment epithelium. Investigative Ophthalmology & Visual Science. 2010;**51**:5314-5319. DOI: 10.1167/iovs.10-5280

[90] Martina JA, Diab HI, Lishu L, Jeong-A L, Patange S, Raben N, Puertollano R. The nutrient-responsive transcription factor TFE3 promotes autophagy, lysosomal biogenesis, and clearance of cellular debris. Science Signaling. 2014;**7**:ra9. DOI: 10.1126/scisignal.2004754

[91] Zhang J, Bai Y, Huang L, Qi Y, Zhang Q, Li S, Wu Y, Li X. Protective effect of autophagy on human retinal pigment epithelial cells against lipofuscin fluorophore A2E: implications for age-related macular degeneration. Cell Death & Disease. 2015;**6**:e1972. DOI: 10.1038/cddis.2015.330

[92] Wong WT, Dresner S, Forooghian F, Glaser T, Doss L, Zhou M, Cunningham D, Shimel K, Harrington M, Hammel K, Cukras CA, Ferris FL, Chew EY. Treatment of geographic atrophy with subconjunctival sirolimus: Results of a phase I/II clinical trial subconjunctival sirolimus for treatment of GA. Investigative Ophthalmology & Visual Science. 2013;**54**:2941-2950

[93] Rodríguez-Muela N, Koga H, García-Ledo L, de la Villa P, de la Rosa EJ, Cuervo AM, Boya P. Balance between autophagic pathways preserves retinal homeostasis. Aging Cell. 2013;**12**:478-488. DOI: 10.1111/acel.12072

[94] De S, Sakmar TP. Interaction of A2E with model membranes. Implications to the pathogenesis of age-related macular degeneration. Journal of General Physiology. 2002;**120**:147-157. DOI: 10.1085/jgp.20028566

[95] Schutt F, Bergmann M, Holz FG, Kopitz J. Isolation of intact lysosomes from human RPE cells and effects of A2-E on the integrity of the lysosomal and other cellular membranes. Graefe's Archive for Clinical and Experimental Ophthalmology. 2002;**240**:983-988. DOI: 10.1007/s00417-002-0558-8

[96] Rajamäki K, Lappalainen J, Oörni K, Välimäki E, Matikainen S, Kovanen PT, Eklund KK, Va E, Matikainen S, Petri T, Rajama K, Eklund KK. Cholesterol crystals activate the NLRP3 inflammasome in human macrophages: A novel link between cholesterol metabolism and inflammation. PLoS One. 2010;**5**:e11765. DOI: 10.1371/journal.pone.0011765

[97] Grebe A, Latz E. cholesterol crystals and inflammation. Current Rheumatology Reports. 2013;**15**:313. DOI: 10.1007/s11926-012-0313-z

[98] Anderson OA, Finkelstein A, Shima DT. A2E induces IL-1ß production in retinal pigment epithelial cells via the NLRP3 inflammasome. PLoS One. 2013;**8**:e67263. DOI: 10.1371/journal.pone.0067263

[99] Haralampus-grynaviski NM, Lamb LE, Clancy CMR, Skumatz C, Burke JM, Sarna T, Simon JD. Spectroscopic and morphological studies of human retinal lipofuscin granules. Proceedings of the National Academy of Sciences. 2003;**100**:3179-3184

[100] Schutt F, Bergmann M, Holz FG, Dithmar S, Volcker HE, Kopitz J. Accumulation of A2-E in mitochondrial membranes of cultured RPE cells. Graefe's Archive for Clinical and Experimental Ophthalmology. 2007;**245**:391-398. DOI: 10.1007/s00417-006-0376-5

[101] Repnik U, Hafner Česen M, Turk B. Lysosomal membrane permeabilization in cell death: Concepts and challenges. Mitochondrion. 2014;**19**:49-57. DOI: 10.1016/j.mito.2014.06.006

[102] Vives-bauza C, Anand M, Shiraz AK, Shirazi AK, Magrane J, Gao J, Vollmer-snarr HR, Manfredi G, Finnemann SC. The age lipid A2E and mitochondrial dysfunction synergistically impair phagocytosis by retinal pigment epithelial cells. Journal of Biological Chemistry. 2008;**283**:24770-24780. DOI: 10.1074/jbc.M800706200

[103] Lakkaraju A, Finnemann SC, Rodriguez-Boulan E. The lipofuscin fluorophore A2E perturbs cholesterol metabolism in retinal pigment epithelial cells. Proceedings of the National Academy of Sciences of the United States of America. 2007;**104**:11026-11031. DOI: 10.1073/pnas.0702504104

[104] Iriyama A, Fujiki R, Inoue Y, Takahashi H, Tamaki YY, Takezawa S, Takeyama K, Jang WD, Kato S, Yanagi Y. A2E, a pigment of the lipofuscin of retinal pigment epithelial cells, is an endogenous ligand for retinoic acid receptor. Journal of Biological Chemistry. 2008;**283**:11947-11953. DOI: 10.1074/jbc.M708989200

[105] Iriyama A, Inoue Y, Takahashi H, Tamaki Y, Jang W, Yanagi Y. A2E, a component of lipofuscin, is pro-angiogenic in vivo. Journal of Cellular Physiology. 2009;**220**:469-475. DOI: 10.1002/jcp.21792

[106] Moiseyev G, Nikolaeva O, Chen Y, Farjo K, Takahashi Y, Ma J. Inhibition of the visual cycle by A2E through direct interaction with RPE65 and implications in Stargardt disease. Proceedings of the National Academy of Sciences of the United States of America. 2010;**107**:17551-17556

[107] Singer JR, Bakall B, Gordon GM, Reddy RK. Treatment of vitamin A deficiency retinopathy with sublingual vitamin A palmitate. Documenta Ophthalmologica 2016;**132**: 137-145. DOI: 10.1007/s10633-016-9533-2

[108] Radu RA, Han Y, Bui TV, Nusinowitz S, Bok D, Lichter J, Widder K, Travis GH, Mata NL. Reductions in serum vitamin A arrest accumulation of toxic retinal fluorophores: A potential therapy for treatment of lipofuscin-based retinal diseases. Investigative Ophthalmology & Visual Science. 2005;**46**:4393-4401

[109] Charbel Issa P, Barnard AR, Herrmann P, Washington I, MacLaren RE. Rescue of the Stargardt phenotype in Abca4 knockout mice through inhibition of vitamin A dimerization. Proceedings of the National Academy of Sciences. 2015;**112**:8415-8420. DOI: 10.1073/pnas.1506960112

[110] Julien S, Schraermeyer U. Lipofuscin can be eliminated from the retinal pigment epithelium of monkeys. Neurobiology of Aging. 2012;**33**:2390-2397. DOI: 10.1016/j.neurobiolaging.2011.12.009

[111] Wu Y, Zhou J, Fishkin N, Rittmann BE, Sparrow JR. Enzymatic degradation of A2E, a retinal pigment epithelial lipofuscin bisretinoid. Journal of the American Chemical Society 2011;133:849-857. DOI: 10.1021/ja107195u

[112] Sparrow JR, Zhou J, Ghosh SK, Liu Z. Bisretinoid degradation and the ubiquitin-proteasome system. Advances in Experimental Medicine and Biology. 2014;**801**:593-600. DOI: 10.1007/978-1-4614-3209-8_75

[113] Yogalingam G, Lee AR, Mackenzie DS, Maures TJ, Rafalko A, Prill H, Berguig G, Hague C, Christianson T, Bell SM, LeBowitz JH. Cellular uptake and delivery of Myeloperoxidase to lysosomes promotes lipofuscin degradation and lysosomal stress in retinal cells. Journal of Biological Chemistry. 2017. DOI: 10.1074/jbc.M116.739441

[114] Stella VJ, He Q. Cyclodextrins. Toxicologic Pathology. 2008;**36**:30-42. DOI: 10.1177/0192623307310945

[115] Nociari MM, Lehmann GL, Perez Bay AE, Radu RA, Jiang Z, Goicochea S, Schreiner R, Warren JD, Shan J, Adam de Beaumais S, Ménand M, Sollogoub M, Maxfield FR, Rodriguez-Boulan E. Beta cyclodextrins bind, stabilize, and remove lipofuscin bisretinoids from retinal pigment epithelium. Proceedings of the National Academy of Sciences of United States of America. 2014;**111**:E1402-E1408. DOI: 10.1073/pnas.1400530111

[116] Maeda A, Maeda T, Golczak M, Palczewski K. Retinopathy in mice induced by disrupted all-trans-retinal clearance. Journal of Biological Chemistry. 2008;**283**:26684-26693. DOI: 10.1074/jbc.M804505200

[117] Mondal M, Mesmin B, Mukherjee S, Maxfield FR. Sterols are mainly in the cytoplasmic leaflet of the plasma membrane and the endocytic recycling compartment in CHO cells. Molecular Biology of the Cell. 2009;**20**:581-588. DOI: 10.1091/mbc.E08-07-0785

[118] López CA, de Vries AH, Marrink SJ. Computational microscopy of cyclodextrin mediated cholesterol extraction from lipid model membranes. Scientific Reports. 2013;**3**:1-6. DOI: 10.1038/srep02071

[119] Vance, Jean E., and Barbara Karten. "Niemann-Pick C Disease and Mobilization of Lysosomal Cholesterol by Cyclodextrin." Journal of Lipid Research 2014;**55**:1609-1621. PMC. Web. 24 May 2017

[120] Yancey PG, Rodrigueza WV, Kilsdonk EPC, Stoudt GW, Johnson WJ, Phillips MC, Rothblat GH. Cellular cholesterol efflux mediated by cyclodextrins: Demonstration of kinetic pools and mechanism of efflux. Journal of Biological Chemistry. 1996;**271**:16026-16034. DOI: 10.1074/jbc.271.27.16026

[121] Rosenbaum AI, Zhang G, Warren JD, Maxfield FR. Endocytosis of beta-cyclodextrins is responsible for cholesterol reduction in Niemann-Pick type C mutant cells. Proceedings of the National Academy of Sciences of the United States of America. 2010;**107**:5477-5482. DOI: 10.1073/pnas.0914309107

[122] Mondjinou YA, McCauliff LA, Kulkarni A, Paul L, Hyun SHH, Zhang Z, Wu Z, Wirth M, Storch J, Thompson DH. Synthesis of 2-hydroxypropyl-β-cyclodextrin/pluronic-based polyrotaxanes via heterogeneous reaction as potential Niemann-Pick type C therapeutics. Biomacromolecules. 2013;**14**:4189-4197. DOI: 10.1021/bm400922a

[123] Chu BB, Liao YC, Qi W, Xie C, Du X, Wang J, Yang H, Miao HH, Li BL, Song BL. Cholesterol transport through lysosome-peroxisome membrane contacts. Cell. 2015;**161**:291-306. DOI: 10.1016/j.cell.2015.02.019

[124] Neufeld EB, Cooney AM, Pitha J, Dawidowicz EA, Dwyer NK, Pentchev PG, Blanchette-Mackie EJ. Intracellular trafficking of cholesterol monitored with a cyclodextrin. Journal of Biological Chemistry. 1996;**271**:21604-21613

[125] Maxfield FR, Wüstner D. Analysis of cholesterol trafficking with fluorescent probes. Methods in Cell Biology. 2012;**108**:367-393. DOI: 10.1016/B978-0-12-386487-1.00017-1

[126] Taylor AM, Liu B, Mari Y, Liu B, Repa JJ. Cyclodextrin mediates rapid changes in lipid balance in Npc1−/− mice without carrying cholesterol through the bloodstream. Journal of Lipid Research. 2012;**53**:2331-2342. DOI: 10.1194/jlr.M028241

[127] Rothblat GH, de la Llera-Moya M, Atger V, Kellner-Weibel G, Williams DL, Phillips MC. Cell cholesterol efflux: Integration of old and new observations provides new insights. Journal of Lipid Research. 1999;**40**:781-796. DOI: 10.1074/jbc.271.27.16026

[128] Ghosh A, Jana M, Modi K, Gonzalez FJ, Sims KB, Berry-Kravis E, Pahan K. Activation of peroxisome proliferator-activated receptor α induces lysosomal biogenesis in brain cells: Implications for lysosomal storage disorders. Journal of Biological Chemistry. 2015;**290**. DOI: 10.1074/jbc.M114.610659

[129] Zimmer S, Grebe A, Bakke SS, Bode N, Halvorsen B, Ulas T, Skjelland M, De Nardo D, Labzin LI, Kerksiek A, Hempel C, Heneka MT, Hawxhurst V, Fitzgerald ML, Trebicka J, Bjorkhem I, Gustafsson JA, Westerterp M, Tall AR, Wright SD, Espevik T, Schultze JL, Nickenig G, Lutjohann D, Latz E. Cyclodextrin promotes atherosclerosis regression via macrophage reprogramming. Science Translational Medicine. 2016;**8**:333ra50. DOI: 10.1126/scitranslmed.aad6100

[130] Yao J, Ho D, Calingasan NY, Pipalia NH, Lin MT, Beal MF. Neuroprotection by cyclodextrin in cell and mouse models of Alzheimer disease. Journal of Experimental Medicine. 2012;**209**:2501-2513. DOI: 10.1084/jem.20121239

[131] Hoh Kam J, Lynch A, Begum R, Cunea A, Jeffery G. Topical cyclodextrin reduces amyloid beta and inflammation improving retinal function in ageing mice. Experimental Eye Research. 2015;**135**:59-66. DOI: 10.1016/j.exer.2015.03.023

[132] Bar-On P, Rockenstein E, Adame A, Ho G, Hashimoto M, Masliah E. Effects of the cholesterol-lowering compound methyl-beta-cyclodextrin in models of alpha-synucleinopathy. Journal of Neurochemistry. 2006;**98**:1032-1045. DOI: 10.1111/j.1471-4159.2006.04017.x

[133] Song W, Wang F, Lotfi P, Sardiello M, Segatori L. 2-Hydroxypropyl-β-cyclodextrin promotes TFEB-mediated activation of autophagy: Implications for therapy. Journal of Biological Chemistry. 2014;**289**:0-27. DOI: 10.1074/jbc.M113.506246

[134] Song W, Wang F, Savini M, Ake A, di Ronza A, Sardiello M, Segatori L. TFEB links autophagy to lysosomal biogenesi. Human Molecular Genetics. 2013;**22**:1994-2009. DOI: 10.1093/hmg/ddt052

[135] Cheng J, Ohsaki Y, Tauchi-Sato K, Fujita A, Fujimoto T. Cholesterol depletion induces autophagy. Biochemical and Biophysical Research Communications. 2006;**351**:246-252. DOI: 10.1016/j.bbrc.2006.10.042

[136] Chen FW, Li C, Ioannou YA. Cyclodextrin induces calcium-dependent lysosomal exocytosis. PLoS One. 2010;**5**:e15054. DOI: 10.1371/journal.pone.0015054

[137] Kilpatrick K, Zeng Y, Hancock T, Segatori L. Genetic and chemical activation of TFEB mediates clearance of aggregated α. Synuclein. 2015:1-21. DOI: 10.1371/journal.pone.0120819

[138] Spampanato C, Feeney E, Li L, Cardone M, Lim JA, Annunziata F, Zare H, Polishchuk R, Puertollano R, Parenti G, Ballabio A, Raben N. Transcription factor EB (TFEB) is a new therapeutic target for Pompe disease. EMBO Molecular Medicine. 2013;**5**:691-706. DOI: 10.1002/emmm.201202176

[139] Tsunemi T, Ashe TD, Morrison BE, Soriano KR, Au J, Roque RAV, Lazarowski ER, Damian VA, Masliah E, La Spada AR. PGC-1α rescues Huntington's disease proteotoxicity by preventing oxidative stress and promoting TFEB function. Science Translational Medicine. 2012;**4**:142ra97. DOI: 10.1126/scitranslmed.3003799

Chapter 2

Lysosomal Dysfunctions in Hereditary Spastic Paraplegias

Haruo Shimazaki

Additional information is available at the end of the chapter

http://dx.doi.org/10.5772/intechopen.70142

Abstract

Hereditary spastic paraplegias (HSPs) comprise a heterogeneous group of inherited neurodegenerative diseases with the cardinal feature of a length-dependent degeneration of corticospinal motor axons. They are classified by their mapped genetic loci, SPG1–SPG78. Recently, lysosomal dysfunction is one of the pathomechanism for some autosomal-recessive HSPs. SPG11 is caused by loss-of-function mutations in the *SPG11* gene. Its gene product is called spatacsin, which is needed for the recycling of lysosomes from autolysosomes. SPG15 is caused by loss-of-function mutation in the *ZFYVE26* gene. The ZFYVE26 gene encodes spastizin. Mutations in spastizin impair autophagosome maturation and lead to an accumulation of immature autophagosomes. *SPG48/KIAA0415* encodes AP5Z1, known to be a spatacsin and spastizin interactor. Its mutations lead to loss of protein or mutated forms of protein with defective autophagy. The TECPR2 is a human ATG8-binding protein and positive regulator of autophagy, which plays a key role in major adult and pediatric neurodegenerative diseases. Mutations in the *lysosomal trafficking regulator* (*LYST*) gene have been reported to cause hereditary spastic paraplegia. The LYST protein is involved in control of the exocytosis of secretory lysosomes. Recently, Drosophila with a gene mutation of an LYST homolog was revealed to exhibit impaired autophagy.

Keywords: SPG11, SPG15, SPG48, SPG48, SPG49, Chediak-Higashi syndrome, lysosomal trafficking regulator, autophagy

1. Introduction

Lysosomal dysfunction, especially disturbance of the autophagy-lysosomal system, substantially contributes to the pathodynamics of some major neurodegenerative disorders: Alzheimer disease, Parkinson disease, Huntington disease, frontotemporal dementia, amyotrophic lateral sclerosis, bulbospinal muscular atrophy and spinocerebellar ataxia 3, hereditary spastic paraplegias (HSPs), and so on [1].

Hereditary spastic paraplegias (HSPs) comprise a heterogeneous group of inherited neurodegenerative diseases with the cardinal feature of a length-dependent degeneration of corticospinal motor axons [2]. They are classified by their mapped genetic loci, SPG1–SPG78. To date, over 60 causative genes have been identified, transmitted by autosomal-dominant, autosomal-recessive (AR), X-linked recessive inheritances, with *de novo* mutations also described.

Several pathogenic mechanisms of HSPs were suggested by the studies in several causative genes for HSP. HSP might result from disruption of the axonal transport of molecules, organelles, and other cargos, which mainly affect the distal parts of motor neurons. Other mechanisms for developing HSP are endoplasmic reticulum formation, membrane trafficking, mitochondrial function [3], lipid metabolism, and myelination. Recently, lysosomal dysfunction is one of the pathomechanism for some autosomal-recessive HSPs: SPG11, SPG15, SPG49, SPG78, and HSP with *lysosomal trafficking regulator* (*LYST*) gene mutation [4, 5].

2. SPG11

The most prevalent autosomal-recessive (AR) HSP is SPG11. SPG11 is characterized by early onset spastic paraplegia with mental impairment and peripheral neuropathy. Brain MRI shows thin corpus callosum (TCC) [6] (**Figure 1**).

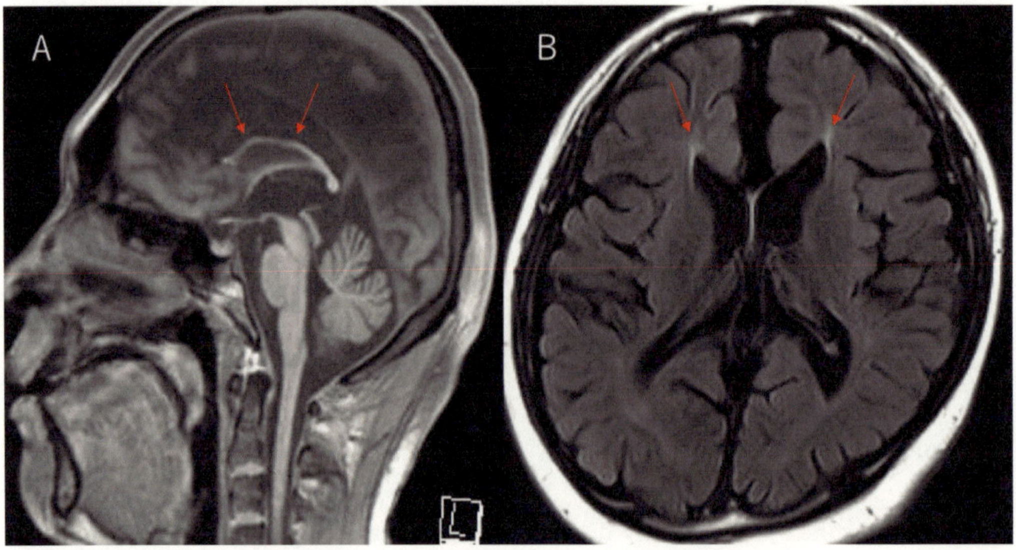

Figure 1. Brain MRI findings of the SPG11 patient. (A) Sagittal T1 image showing a thin corpus callosum, especially the anterior part. (B) Axial FLAIR image showed subtle ears of the lynx formation at the anterior periventricular portions of the lateral ventricles.

There may be upper limb involvement and nystagmus. Some patients additionally present with parkinsonism or ataxia. Patients with an amyotrophic lateral sclerosis (ALS)-like upper motor neuron-dominant phenotype but with atypically long disease duration and absence of TCC or mental impairment (ALS5) had been reported [7]. Moreover, peripheral neuropathy-dominant phenotype (Charcot-Marie-Tooth disease like) has been described recently [8].

Pathological features of SPG11 in neurons are mainly intracytoplasmic granular lysosome-like structures in supratentorial areas, and others in subtentorial areas are ubiquitin and p62 aggregates, observed in amyotrophic lateral sclerosis (ALS), except that they are never labeled with anti-TDP-43 or anti-cystatin C [9].

SPG11 is caused by loss-of-function mutations in the *KIAA1840* gene on chromosome 15q. *KIAA1840*, encoding spatacsin, is expressed ubiquitously in the nervous system, but most prominently in the cerebellum, cerebral cortex, hippocampus, and pineal gland. Its gene product is called spatacsin, which is needed for the recycling of lysosomes from autolysosomes, a process known as autophagic lysosome reformation (ALR). Loss of spatacsin (or spastizin) resulted in the depletion of free lysosomes, which are competent to fuse with autophagosomes, and an accumulation of autolysosomes, reflecting a failure in ALR. Moreover, spatacsin and spastizin were essential components for the initiation of lysosomal tubulation [10]. Fibroblasts prepared from patients with SPG11 have selective enlargement of LAMP1-positive structures, though prominent abnormal lysosomal storage was not evident [11].

In spatacsin-null mice, lysosomes are diminished in cortical neurons and Purkinje cells *in vivo*. The decreased number of lysosomes useful for fusion with autophagosomes damages autolysosomal clearance, results in the accumulation of undegraded substances, and finally causes death of particularly susceptible neurons like cortical motor neurons and Purkinje cells in knockout mice [12].

3. SPG15

SPG15 is the second most common type of hereditary spastic paraplegia with thin corpus callosum. SPG15 cases show similar symptoms to those observed in SPG11. Clinical manifestation in addition to spastic paraplegia includes intellectual disability, pigmentary retinopathy (Kjellin syndrome), cerebellar ataxia, parkinsonism, and axonal neuropathy. Brain MRI often shows thin corpus callosum.

SPG15, the second most autosomal-recessive HSP, is caused by loss-of-function mutation in the *ZFYVE26/SPG15* gene. The ZFYVE26 gene encodes a large protein of 2539 amino acid residues termed spastizin.

In zebrafish, spastizin, together with spatacsin, is essential for proper establishment of the motor neuron axonal network; these proteins are indeed necessary for outgrowth and proper targeting of motor neuron axons [13].

Spastizin interacts with the autophagy-related Beclin 1-UVRAG-Rubicon multiprotein complex and is involved in autophagosome maturation. Mutations in spastizin disrupt its interaction with Beclin 1 and thus with the complex, damaging autophagosome maturation and resulting in a retention of immature autophagosomes in patient's fibroblasts. Similarly, a deposition of autophagosomes was detected in SHSY5Y cells and in primary hippocampal neurons after spastizin silencing, thus indicating that autophagy impairments by spastizin deficiency give rise to both neuronal and non-neuronal cells [14].

Spastizin is expressed in the identical organs as spatacsin (SPG11) and partially coexists with microtubules, mitochondria, and the nucleus. Spastizin is also observed at the midbody during cytokinesis. The spastizin interacts with spatacsin and with KIAA0415 (SPG48), a member of the AP5 complex. Mutations lead to loss of protein or mutated forms of protein with defective autophagy [14].

Zfyve26 knockout mice developed normally by 12 months of age acquire a spastic and ataxic gait disorder accompanied by neuron loss in the motor cortex and the cerebellum, consistent with the clinical phenotype of SPG15 patients [15]. High-density LAMP1-positive membrane-bound vesicles and lipopigment accumulate in neurons of *Zfyve26* knockout mice [15].

4. SPG48

Two siblings in the one French family have pure adult-onset spastic paraplegia and urinary incontinence with hyperintensity of the cervical spinal cord (C3-4,7) in one sibling as the only distinguishing magnetic resonance imaging (MRI) feature [16]. Thereafter, clinical features included not only prominent spastic paraparesis but also sensory and motor neuropathy, ataxia, dystonia, parkinsonism, and myoclonus. Skin fibroblasts from SPG48 patients tested positive for periodic acid Schiff (PAS) and intrinsic fluorescence material, while electron microscopic analysis indicated lamellar material concordant with abnormal storage of lysosomal material [17].

An insertion/deletion mutation has been identified in *SPG48/KIAA0415* in two HSP families. *KIAA0415* encodes a presumptively helicase (AP5Z1), localized in both nucleus and cytoplasm, participated in DNA double-strand break repair processes, and interacted with spatacsin and spastizin. Especially, spatacsin is phosphorylated upon DNA damage by protein kinases ataxia telangiectasia mutated (ATM) or ATR (ATM and Rad3-related) [16]. A recent study showed that the protein is a member of the adaptor protein 5 complex (AP5) that is implicated in vesicle formation and sorting (as AP4) [18].

5. SPG49

A homozygous truncating mutation in SPG49/TECPR2 (tectonin beta-propeller repeat containing 2) was identified in a new form of complicated HSP [19]. This HSP is characterized by early onset of spastic paraplegia, motor development delaying, mental retardation, dysmorphic features (short stature, round face, low anterior hairline, dental crowding, short broad neck, mild brachycephalic microcephaly, a chubby appearance), cerebellar dysarthria, ataxia, episodes of central apnea, and TCC on brain MRI.

TECPR2 protein has been established to be a binding partner of the mammalian Atg8 protein family, including LC3, and a probable positive regulator of autophagosome formation [20]. Using fibroblast of affected SPG49 patients and knockdown of TECPR2 using siRNA in cultured cell lines, loss of TECPR2 was found to result in a decreased number of autophagosomes and reduced delivery of LC3 and p62 for lysosomal degradation [19]. Recent study showed that TECPR2 is involved in maintaining functional endoplasmic reticulum exit sites, which may serve as scaffolds for the formation of autophagosomes [21].

6. SPG78

The disease presentation in SPG78 patients was dominated by an adult-onset lower-limb predominant spastic paraparesis. Cognitive impairment was present in most of the cases and ranged from very mild deficits to advanced dementia with frontotemporal characteristics. Nerve conduction studies revealed involvement of the peripheral motor and sensory nerves. Only one of five patients with hereditary spastic paraplegia showed clinical indication of extrapyramidal involvement in the form of subtle bradykinesia and slight resting tremor. Neuroimaging cranial investigations revealed pronounced vermian and hemispheric cerebellar atrophy. Notably, reduced striatal dopamine was apparent in the brain of one of the patients, who had no clinical signs or symptoms of extrapyramidal involvement [22].

ATP13A2, which is causative protein of SPG78, is a lysosomal P5-type transport ATPase, the activity of which critically depends on catalytic autophosphorylation. Biochemical and immunocytochemical experiments in COS-1 and HeLa cells and SPG78 patient-derived fibroblasts demonstrated that the hereditary spastic paraplegia-associated mutations, similarly to the ones causing Kufor-Rakeb syndrome and neuronal ceroid lipofuscinosis, cause loss of ATP13A2 function due to transcript or protein instability and abnormal intracellular localization of the mutant proteins, ultimately impairing the lysosomal and mitochondrial function [22]. They confirm in fibroblast of SPG78 patients that LAMP1-positive organelles accumulate, correlating with a reduction in their proteolytic activity. These findings are similar to that of SPG11 and SPG15 [10, 11, 14]. ATP13A2 has been implicated in autophagy pathway [23].

7. HSP due to *LYST* mutation

We encountered an autosomal-recessive (AR) HSP family with cerebellar ataxia and neuropathy whose gene locus was not linked to previously reported AR-HSP loci. We have identified a novel homozygous missense mutation in the *lysosomal trafficking regulator* (*LYST*) gene, a nuclear gene encoding for a protein involved in intracellular trafficking [24]. *LYST* is described as the causative gene for Chediak-Higashi syndrome (CHS, OMIM #214500), which is a rare autosomal-recessive syndrome characterized by hypopigmentation, severe immune-deficiency, a bleeding tendency, and progressive neurological dysfunction [25]. We describe a CHS family including two patients who show variable degrees of spastic paraplegia, cerebellar ataxia, and neuropathy, whereas they exhibit no apparent skin hypopigmentation and blood or immune system abnormalities (**Figure 2**).

The study included two patients in a Japanese consanguineous family (**Figure 2**). Neurologic examination was performed in two patients. Detailed clinical and laboratory findings of the two patients were shown in **Table 1** and **Figures 3** and **4**. Blood examinations revealed peroxidase-positive giant granules in granulocytes in two cases (**Figure 4**).

DNA analysis was done in two patients and two normal family members. We performed a genomewide linkage analysis employing SNP arrays with two patients' DNAs (**Figure 5A**) and exome sequencing using one patient's sample. We confirmed the mutation by Sanger sequencing of the family members (**Figure 5B**).

We identified a homozygous missense mutation (c.4189T>G, p.F1397V) in the *lysosomal trafficking regulator* (*LYST*) gene in the two patients. (**Figure 5B**). This mutation co-segregated with the disease in the family and located at the well-conserved amino acid. (**Figure 5C**). We also detected a heterozygous nonsense mutation (c. 823C>T, p.R275X) in the *BSCL2* gene in one patient (III-3), not in another one (III-1) (**Figure 5C**). *BSCL2* is the causative gene for SPG17. This mutation is deleterious [26], and we suggest that this nonsense mutation is implicated in phenotype differences between two cases: leg spasticity of III-3 is stronger than that of III-1.

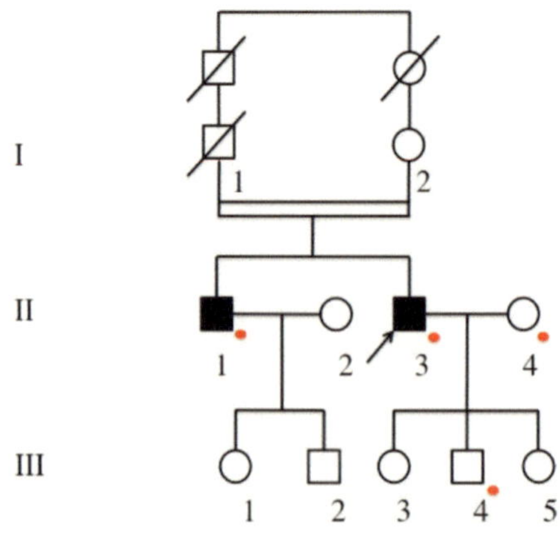

Figure 2. Family pedigree with *LYST* gene mutation. The proband (III-3) and his older brother (III-1) show the almost same clinical phenotype. Other members are all healthy except for deceased elderly ones. The parents (II-1 and 2), who were first cousins, were neurologically asymptomatic. Arrow indicates the proband, and dots indicate the persons who permit analysis of their DNAs.

	III-1	III-3
Age at examination	63	53
Onset	58	48
Leg spasticity	–	+
Limb ataxia	+	+
Thigh muscle atrophy	+	+
Patellar Tendon Reflex (PTR)	–	++
Achilles Tendon Reflex (ATR)	+ –	–
Babinski sign	+	+
Vibration sense	Normal	Normal
Sphincter involvement	–	–
Mini-Mental State Examination (MMSE)	16/30	25/30
Lower limb NCV	FWCV decreased	Decreased
Sural nerve biopsy	Axonal swelling, myelin↓	(Not examined)
Brain MRI	Cerebellar atrophy	Cerebellar atrophy
Spinal MRI	Normal	Thoracic atrophy
Large granules in granulocytes	+	+
Phagocytic activity of leukocytes	98.7%	(Not examined)
NK-cell activity (18–40)	11	14

Table 1. Clinical and laboratory presentation of the two patients.

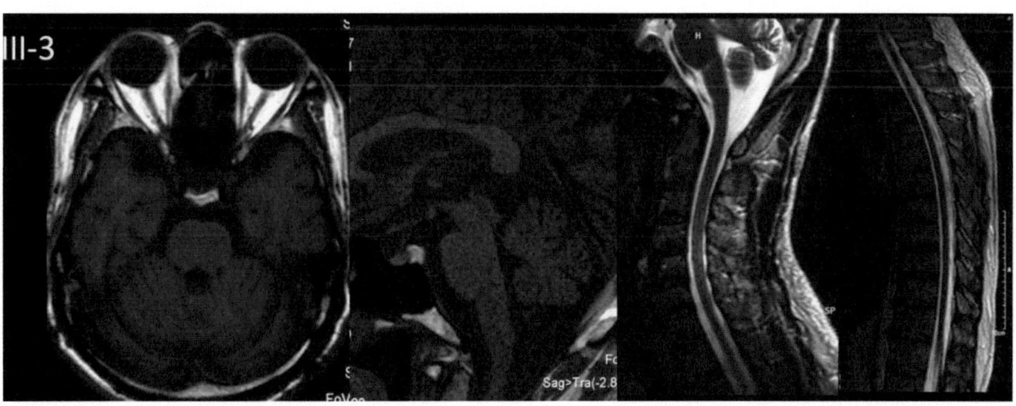

Figure 3. Brain and spinal MRI of the patient III-3. The brain MRI showed mild cerebellar atrophy, and spinal MRI disclosed mild thoracic cord atrophy.

The gene responsible for CHS was identified in 1996 and called *lysosomal trafficking regulator* (*LYST*) [27, 28]. The *LYST* gene is a large gene that has 51 coding exons and an open reading frame (ORF) of 11,403 kb [6]. The LYST protein, which is a large, putative cytosolic protein of

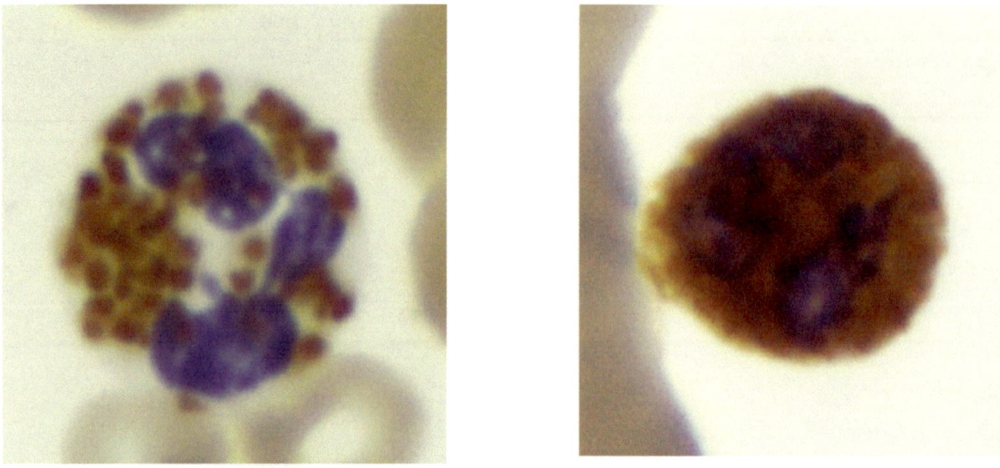

Figure 4. Peripheral blood leukocyte of patient III-3 (peroxidase stain). We found peroxidase-positive large granules in the patient's granulocytes compared with normal control WBC. These patients had no symptoms according to immunodeficiency or bleeding tendency.

425 kDa (3801 amino acids), is ubiquitously expressed and involved in control of the exocytosis of secretory lysosomes [28, 29]. The LYST protein has a BEACH (named after BEige And Chediak-Higashi) domain (amino acid numbers 3132–3422) [28], Trp-Asp (WD) 40 repeats (amino acid numbers 3477–3778), and a concanavalin (Con)A-like lectin domain (amino acid numbers 1390–1691) [30]. The LYST protein has been proposed to act as a scaffold protein in the mediation of fusion or a fission event of vesicles [31]. The mutation in this family (p.F1397V) is located within the ConA-like lectin domain. This domain could be involved in oligosaccharide binding associated with protein traffic and sorting along the secretory pathway [30].

Dysfunctional secretion of enlarged lysosome-related organelles, including lysosomes, melanosomes, and cytolytic granules, has been observed in cells with mutations in *LYST*. Small interfering RNA knockdown of LYST in human cell lines replicates the *LYST*-mutant phenotype of large lysosomes [32]. They found no evidence that autophagy or endocytic degradation was affected by LYST depletion. Autophagosomes are formed in normal size and volume and are able to form the large fused lysosomes, resulting in normal degradation rates. The large lysosomes are fully functioned in degrading endogenous proteins. LYST did not affect retrograde trafficking of toxins as well as the localization of transporters of lysosomal proteins, adaptor protein-3 (AP-3), and cation-independent mannose-6-phosphate receptor (CI-MPR). The large lysosomes quantitative analysis demonstrates that LYST depletion results in reduction in vesicle content per cell; meanwhile, the total enzymatic amount and vesicular pH are unaffected, indicating a role for LYST in lysosomal fission and/or fusion events [32].

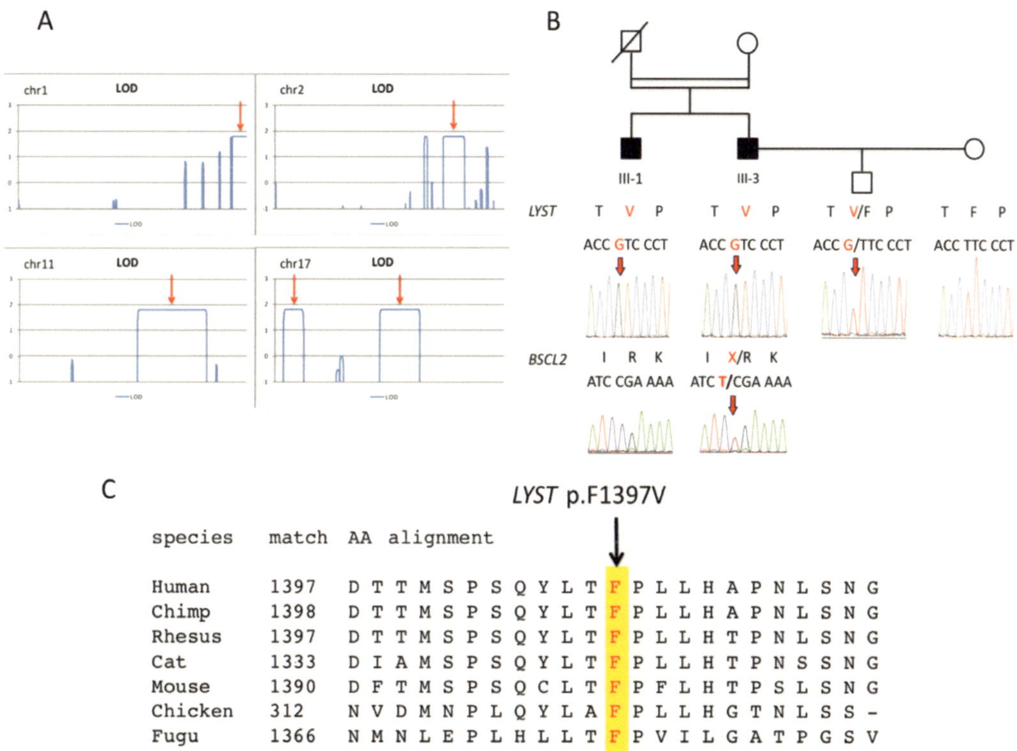

Figure 5. Linkage analysis and mutation of the *lysosomal trafficking regulator* (*LYST*) gene in the patients. (A) Linkage analysis. Linkage analysis involving SNPs revealed the highest LOD scores (about 1.8) in parts of chromosomes 1, 2, 11, and 17 (arrows). These four areas were thought to be candidate areas in which the causative gene was located. (B) Sanger sequencing confirmed the homozygous nonsense mutation (c.4189T>G, p.F1397V) of the LYST gene identified in the proband (III-3) and the affected brother (III-1). This mutation co-segregated with the disease in this family. This *LYST* mutation was not found in 200 Japanese control DNAs. In one patient (III-3), we identified a heterozygous nonsense mutation (c.823C>T, p.R275*) of the *BSCL2* gene, the causative one for SPG17, whereas no mutation in III-1. (C) This mutation located at the highly conserved residues within the BEACH (named after BEige And Chediak-Higashi) and concanavalin A (ConA)-like lectin domain.

Nevertheless, Drosophila with gene mutation of *LYST* homolog revealed impaired autophagy [33]. The roles of LYST in autophagy remain controversial [34].

Author details

Haruo Shimazaki

Address all correspondence to: hshimaza@jichi.ac.jp

Division of Neurology, Department of Internal Medicine, Jichi Medical University, Tochigi, Japan

References

[1] Kenney DL, Benarroch EE. The autophagy-lysosomal pathway: General concepts and clinical implications. Neurology. 2015;**85**(7):634-645

[2] Blackstone C. Cellular pathways of hereditary spastic paraplegia. Annual Review of Neuroscience. 2012;**35**:25-47

[3] Shimazaki H, Takiyama Y, Ishiura H, Sakai C, Matsushima Y, Hatakeyama H, et al. A homozygous mutation of C12orf65 causes spastic paraplegia with optic atrophy and neuropathy (SPG55). Journal of Medical Genetics. 2012;**49**(12):777-784

[4] Lo Giudice T, Lombardi F, Santorelli FM, Kawarai T, Orlacchio A. Hereditary spastic paraplegia: Clinical-genetic characteristics and evolving molecular mechanisms. Experimental Neurology. 2014;**261**:518-539

[5] Noreau A, Dion PA, Rouleau GA. Molecular aspects of hereditary spastic paraplegia. Experimental Cell Research. 2014;**325**(1):18-26

[6] Shimazaki H, Matsuura T. Novel compound heterozygous mutations of SPG11 gene in sporadic spastic paraplegia with thin corpus callosum. In: Proceedings of the 13th International Congress of Human Genetics, p.267; Kyoto. 2016.4.6

[7] Orlacchio A, Babalini C, Borreca A, Patrono C, Massa R, Basaran S, et al. SPATACSIN mutations cause autosomal recessive juvenile amyotrophic lateral sclerosis. Brain: A Journal of Neurology. 2010;**133**(Pt 2):591-598

[8] Montecchiani C, Pedace L, Lo Giudice T, Casella A, Mearini M, Gaudiello F, et al. ALS5/SPG11/KIAA1840 mutations cause autosomal recessive axonal Charcot-Marie-Tooth disease. Brain: A Journal of Neurology. 2016;**139**(Pt 1):73-85

[9] Denora PS, Smets K, Zolfanelli F, Ceuterick-de Groote C, Casali C, Deconinck T, et al. Motor neuron degeneration in spastic paraplegia 11 mimics amyotrophic lateral sclerosis lesions. Brain: A Journal of Neurology. 2016;**139**(Pt 6):1723-1734

[10] Chang J, Lee S, Blackstone C. Spastic paraplegia proteins spastizin and spatacsin mediate autophagic lysosome reformation. The Journal of Clinical Investigation. 2014;**124**(12):5249-5262

[11] Renvoise B, Chang J, Singh R, Yonekawa S, FitzGibbon EJ, Mankodi A, et al. Lysosomal abnormalities in hereditary spastic paraplegia types SPG15 and SPG11. Annals of Clinical and Translational Neurology. 2014;**1**(6):379-389

[12] Varga RE, Khundadze M, Damme M, Nietzsche S, Hoffmann B, Stauber T, et al. In vivo evidence for lysosome depletion and impaired autophagic clearance in hereditary spastic paraplegia type SPG11. PLoS Genetics. 2015;**11**(8):e1005454

[13] Martin E, Yanicostas C, Rastetter A, Alavi Naini SM, Maouedj A, Kabashi E, et al. Spatacsin and spastizin act in the same pathway required for proper spinal motor neuron axon outgrowth in zebrafish. Neurobiology of Disease. 2012;**48**(3):299-308

[14] Vantaggiato C, Crimella C, Airoldi G, Polishchuk R, Bonato S, Brighina E, et al. Defective autophagy in spastizin mutated patients with hereditary spastic paraparesis type 15. Brain: A Journal of Neurology. 2013;**136**(Pt 10):3119-3139

[15] Khundadze M, Kollmann K, Koch N, Biskup C, Nietzsche S, Zimmer G, et al. A hereditary spastic paraplegia mouse model supports a role of ZFYVE26/SPASTIZIN for the endolysosomal system. PLoS Genetics. 2013;**9**(12):e1003988

[16] Slabicki M, Theis M, Krastev DB, Samsonov S, Mundwiller E, Junqueira M, et al. A genome-scale DNA repair RNAi screen identifies SPG48 as a novel gene associated with hereditary spastic paraplegia. PLOS Biology. 2010;**8**(6):e1000408

[17] Hirst J, Madeo M, Smets K, Edgar JR, Schols L, Li J, et al. Complicated spastic paraplegia in patients with AP5Z1 mutations (SPG48). Neurology Genetics. 2016;**2**(5):e98

[18] Hirst J, Barlow LD, Francisco GC, Sahlender DA, Seaman MN, Dacks JB, et al. The fifth adaptor protein complex. PLOS Biology. 2011;**9**(10):e1001170

[19] Oz-Levi D, Ben-Zeev B, Ruzzo EK, Hitomi Y, Gelman A, Pelak K, et al. Mutation in TECPR2 reveals a role for autophagy in hereditary spastic paraparesis. The American Journal of Human Genetics. 2012;**91**(6):1065-1072

[20] Behrends C, Sowa ME, Gygi SP, Harper JW. Network organization of the human autophagy system. Nature. 2010;**466**(7302):68-76

[21] Stadel D, Millarte V, Tillmann KD, Huber J, Tamin-Yecheskel BC, Akutsu M, et al. TECPR2 cooperates with LC3C to regulate COPII-dependent ER export. Molecular Cell. 2015;**60**(1):89-104

[22] Estrada-Cuzcano A, Martin S, Chamova T, Synofzik M, Timmann D, Holemans T, et al. Loss-of-function mutations in the ATP13A2/PARK9 gene cause complicated hereditary spastic paraplegia (SPG78). Brain: A Journal of Neurology. 2017;**140**(Pt 2):287-305

[23] Dehay B, Ramirez A, Martinez-Vicente M, Perier C, Canron MH, Doudnikoff E, et al. Loss of P-type ATPase ATP13A2/PARK9 function induces general lysosomal deficiency and leads to Parkinson disease neurodegeneration. Proceedings of the National Academy of Sciences of the United States of America. 2012;**109**(24):9611-9616

[24] Shimazaki H, Honda J, Naoi T, Namekawa M, Nakano I, Yazaki M, et al. Autosomal-recessive complicated spastic paraplegia with a novel lysosomal trafficking regulator gene mutation. Journal of Neurology, Neurosurgery, and Psychiatry. 2014;**85**(9):1024-1028

[25] Shimazaki H. LYST gene mutations: Mechanisms of Chédiak-Higashi syndrome. The Application of Clinical Genetics. 2017;**10**

[26] Mori E, Fujikura J, Noguchi M, Nakao K, Matsubara M, Sone M, et al. Impaired adipogenic capacity in induced pluripotent stem cells from lipodystrophic patients with BSCL2 mutations. Metabolism. 2016;**65**(4):543-556

[27] Barbosa MD, Nguyen QA, Tchernev VT, Ashley JA, Detter JC, Blaydes SM, et al. Identification of the homologous beige and Chediak-Higashi syndrome genes. Nature. 1996;**382**(6588):262-265

[28] Nagle DL, Karim MA, Woolf EA, Holmgren L, Bork P, Misumi DJ, et al. Identification and mutation analysis of the complete gene for Chediak-Higashi syndrome. Nature Genetics. 1996;**14**(3):307-311

[29] Huynh C, Roth D, Ward DM, Kaplan J, Andrews NW. Defective lysosomal exocytosis and plasma membrane repair in Chediak-Higashi/beige cells. Proceedings of the National Academy of Sciences of the United States of America. 2004;**101**(48):16795-16800

[30] Burgess A, Mornon JP, de Saint-Basile G, Callebaut I. A concanavalin A-like lectin domain in the CHS1/LYST protein, shared by members of the BEACH family. Bioinformatics. 2009;**25**(10):1219-1222

[31] Tchernev VT, Mansfield TA, Giot L, Kumar AM, Nandabalan K, Li Y, et al. The Chediak-Higashi protein interacts with SNARE complex and signal transduction proteins. Molecular Medicine. 2002;**8**(1):56-64

[32] Holland P, Torgersen ML, Sandvig K, Simonsen A. LYST affects lysosome size and quantity, but not trafficking or degradation through autophagy or endocytosis. Traffic. 2014;**15**(12):1390-1405

[33] Rahman M, Haberman A, Tracy C, Ray S, Kramer H. Drosophila mauve mutants reveal a role of LYST homologs late in the maturation of phagosomes and autophagosomes. Traffic. 2012;**13**(12):1680-1692

[34] Cullinane AR, Schaffer AA, Huizing M. The BEACH is hot: A LYST of emerging roles for BEACH-domain containing proteins in human disease. Traffic. 2013;**14**(7):749-766

Chapter 3

Lysosomes: How Plasma Membrane Repair Route Can Be Hijacked by Parasites?

Barbara Hissa and Luciana O. Andrade

Additional information is available at the end of the chapter

http://dx.doi.org/10.5772/intechopen.69305

Abstract

Lysosomes are acidic organelles that are not only involved in degradation processes but also participated in other cellular functions, such as specialized secretion and plasma membrane (PM) resealing. When the PM is ruptured, Ca^{2+} flows from the extracellular milieu toward the cytoplasm potentially triggering cell death. In order to escape from the apoptotic route, cells developed an elegant mechanism in which lysosomes are recruited to the sites of injuries in a Ca^{2+}-dependent fashion. Lysosomes, fuse with the PM releasing their enzymatic content. Acid sphingomyelinase (ASM), one of the secreted enzymes, cleaves sphingomyelin into ceramide, inducing compensatory endocytosis and internalization of the membrane-damaged site. *Trypanosoma cruzi*, the etiological agent of Chagas disease, relies heavily on lysosomes to successfully invade mammalian cells. By mechanically injuring the host PM, *T. cruzi* evokes lysosome exocytosis, and subsequently, compensatory endocytosis. The latter drives the parasite into the host cell, where it can replicate. This early association with lysosomes prevents *T. cruzi* evasion from the host cells allowing colonization of host intracellular milieu. This review chapter will summarize the main contributions in the field exploring the crosstalk between PM repair and *T. cruzi* invasion and how the understanding of these mechanisms evolved throughout the years.

Keywords: plasma membrane repair, lysosomes, exocytosis, compensatory endocytosis, *Trypanosoma cruzi*

1. Introduction

The word lysosome is derived from the Greek words *lysis* (loosening, breaking) and *soma* (body) and literally means 'digestive body'. Those acidic organelles were identified primarily by the biochemist Christian de Duve, in 1955, when he was studying the carbohydrate metabolism, the mechanism of insulin in the liver and the role of an enzyme, known at that time, as hexose phosphate (and later denominated glucose-6-phosphatase) [1]. By doing sucrose gradient centrifugal fractionation, de Duve identified four main fractions on the liver homogenate: nuclear, large granules (mostly composed by mitochondria), small granules (microsomes) and a supernatant. The glucose-6-phosphate enzyme was identified in the microsome fraction [2]. Based on biochemistry enzymatic analysis, de Duve and his group postulated that acid phosphatase must be enclosed within membranous vesicles in such a way that the enzyme could not leak out, and the substrate could not get in [3]. The first morphological observation of a lysosome was performed in 1956, when it was seen under an electron microscope by Novikoff, who later developed the acid phosphatase staining for identifying lysosomes morphologically [4].

Up to this date, more than 50 different enzymes were identified within lysosomes. Those membrane-delimited organelles are present in most nucleated mammalian cells. Lysosomes are mostly scattered across the cytoplasm but can become more concentrated around the perinuclear region upon stimuli [5]. Lysosome intracellular movement is required for its proper functioning and has shown to be tightly regulated in the cell [5]. Given their acidic interior, mostly composed by hydrolases, lysosomes are pivotal in intracellular degradation processes [6] such as intracellular digestion and autophagy [7, 8]. In order to digest endocytic cargo (membrane-bound vesicles resultant from pinocytosis or phagocytosis events) or autophagosomes, lysosomes have to fuse with those vesicles so their enzymes can have access to their content [9–11].

Besides being pivotal for intracellular degradation processes, lysosomes are also important for a plethora of physiological processes inside the cell, such as bone matrix resorption by osteoclasts [12], m-TOR-dependent antigen presentation by macrophages and dendritic cells [13], cholesterol transport [14], Ca^{2+}-regulated PM resealing upon injury [15] and cell death [16], just to cite a few examples. Perturbations in lysosomal homeostasis, such as dysfunction of lysosomal hydrolases, impairment in lysosomal traffic and biogenesis might induce lysosomal storage disorders due to accumulation of unprocessed substrata inside this organelle. There are more than 50 different types of lysosomal storage diseases that were already identified [17].

As mentioned before, lysosomes play an important role in membrane resealing upon injury, and they are a fundamental part of the endocytic pathway. The endocytic pathway is basically composed by early and late endosomes and lysosomes. Internalized particles are delivered to early endosomes and are either recycled back to the membrane or transported to late endosomes. When they reach the late endosomes, the endocytosed material can be sorted by the Golgi apparatus and transported to the membrane or fuse with lysosomes to be degraded [18].

There is no doubt that the endocytic pathway is fundamental for nutrient uptake, cell signalling [19], and migration [20]. A summary of the diverse cellular functions that the lysosomes are involved in is depicted in **Figure 1**. Intriguingly, the endocytic route is also explored by pathogens in order to successfully invade their host cells [21]. Some of these pathogens evolved in order to develop mechanisms to evade lysosomal fusion in order to protect them from being degraded from lysosomal enzymes. However, in some cases, the pathogen drives itself to encounter lysosomes in order to guarantee intracellular survival. The gram-positive bacteria, *Coxiella burnetti*, causative agent of Q fever [22], requires fusion with lysosomes in order to mature its parasitophorous vacuole [PV) and to replicate [23–27]. This pathogen also

http://dx.doi.org/10.5772/intechopen.69305

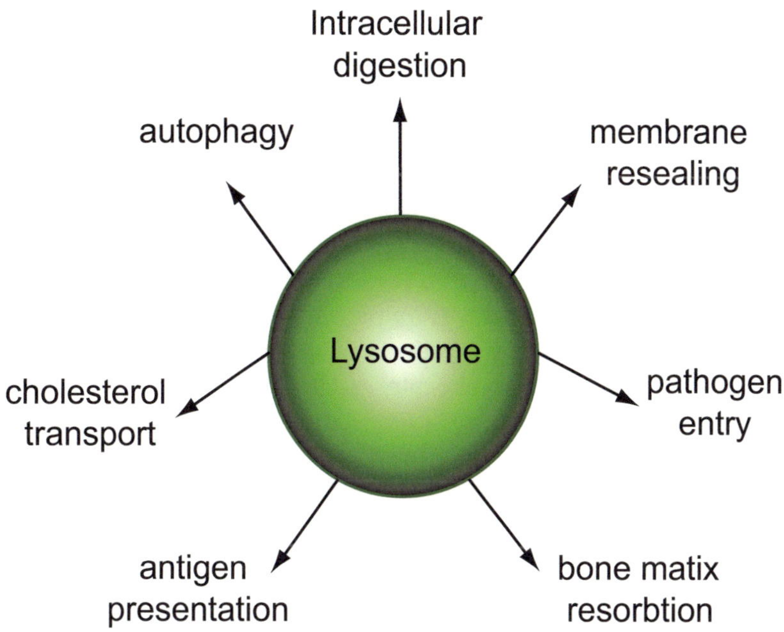

Figure 1. The multitask lysosome. Besides being involved in intracellular digestion, lysosomes also participate in other important cellular functions, such as autophagy, cholesterol transport, antigen presentation, bone matrix resorption, plasma membrane resealing upon injury and pathogen entry in host cells.

requires two lysosomal membrane proteins: lysosomal associated membrane protein 1 and 2 (LAMP-1 and LAMP-2) [28] in order to have normal PV size and bacterial replication rate [29]. Another example is the protozoan parasite *Leishmania donovani* that causes visceral leishmaniasis in humans [30]. By examining the infection of bone marrow–derived macrophages by *L. donovani* metacyclic-derived promastigotes, Forestier and colleagues (2011) demonstrated that the early invasion process is constituted by four phases: (1) contact between highly motile and polarized promastigotes and the PM preceding phagocytosis, (2) formation of the PV, differentiation into amastigote form and intracellular orientation of the parasite, (3) movement of the parasite towards the cell membrane leading to local wounding and (4) PV translocates to the perinuclear region of the host cell. They identified LAMP-1 positive tight PVs as early as 30 min post-infection and demonstrated that the parasites remained viable at those harsh and acidic conditions. Interestingly, at phase 3, when the parasite moves back towards the membrane, it causes membrane rupture evoking lysosomal exocytosis to reseal the membrane. They also found LAMP-1 decorating the parasite's flagellum facing the PM wounded area [31]. The gram-negative bacteria, *Neisseria meningitidis* and *Neisseria gonorrhoeae*, causative agents of meningitis and gonorrhoea, respectively, have a very interesting mechanism of invading mucosal cells. The invasion process can be divided into 4 different steps: (1) attachment, (2) phagocytosis of the bacteria by the host cells at the apical portion, (3) transport of those phagocytosed bacteria to the basal part of the cell and (4) exocytosis of the bacteria-containing vesicles to subepithelial tissues [32]. *Neisseria* secretes an immunoglobulin called IgA1. This immunoglobulin is able to cleave LAMP-1 when this protein is at the host cell

PM. In order to do that this bacteria induces Ca^{2+}-dependent lysosomal exocytosis, and when LAMP-1 is exposed at the surface, it is cleaved by IgA1 [33]. By cleaving LAMP-1, the bacterium also alters other lysosomal constituents, such as LAMP-2, lysosomal acid phosphatase and CD63 [34], which is thought to improve the bacteria intracellular survival.

One of the most interesting pathogens that interact with lysosomes in order to successfully invade host cells is the protozoan parasite *Trypanosoma cruzi. T. cruzi* is the causative agent of Chagas's disease [35], a tropical neglected disease that has no effective vaccine or cure and still affects about 6–7 million people worldwide [36, 37]. Virtually, *T. cruzi* can infect basically all nucleated cells from its mammalian host. The early entry process is complex and involves a plethora of receptors and proteins that are secreted in order to orchestrate parasite attachment and invasion [38]. One step that is pivotal for the parasite entry and infection is its early association with lysosomes [39]. *T. cruzi* subverts the PM wound healing route in order to get access to the intracellular milieu [40]. In fact, the understanding of the intertwined PM repair and *T. cruzi* entry processes evolved in parallel and was elegantly explored by Dr. Norma Andrews' group since the 90s. This chapter will try to explore how the understanding of those mechanisms evolved through time and which are the key players in both membrane healing and *T. cruzi* entry process.

2. From membrane resealing to *Trypanosoma cruzi* invasion: what is the role played by lysosomes?

2.1. Plasma membrane injury and resealing: lysosomes save the day

It has been known since the early 90s that professional secretory cells, such as hepatocytes [41, 42], activated platelets [43, 44], pancreatic acinar cells [45, 46], macrophages [47, 48], osteoclasts [49, 50] and neutrophils [51, 52] are able to undergo regulated lysosomal secretion. However, until the mid-90s, it was not known whether non-professional secretory cells had the capability of performing lysosomal exocytosis. In 1995, Miyake and McNeil demonstrated for the first time that endothelial cells were able to accumulate vesicles near PM injured sites, and those vesicles underwent Ca^{2+} mediated exocytosis in order to seal those wounds [53]. In 1996, Coorssen and colleagues have shown that epithelial cells enlarged their surface area by ~20–30% due to exocytosis promoted by increase in intracellular Ca^{2+}. However, back then, they just hypothesized that the increase in area was probably due to secretion of endosomes or lysosomes [54]. In 1997, Rodriguez and collaborators demonstrated that non-secretory cells, such as fibroblasts, myoblasts and epithelial cells, were able to trigger lysosomal exocytosis upon increase in intracellular Ca^{2+} levels. By performing enzymatic assays, they measured the presence of lysosomal enzymes, such as β-hexosaminidase and cathepsin D, in the supernatant of stimulated cells. In parallel, they also showed the presence of a lysosomal glycoprotein, Igp120, at the PM, corroborating the lysosomal exocytosis hypothesis [55].

Cells have evolved throughout time in order to develop a mechanism by which injuries in the PM could be quickly sealed in order to prevent cytoplasm leakage and cell death. Collagen matrix contraction assays for mimicking tissue morphogenesis and wound healing show that, upon contraction, fibroblasts can uptake extracellular dyes due to the formation of small pores in the membrane. Those small wounds are sealed within 5 s in the presence of Ca^{2+} [56]. Tissues that are under mechanical stress, such as skeletal muscle [57], heart [58], gut [59] and skin [60] also have the ability to reseal their torn membranes and depend on this process for proper functioning. Impairment in sarcolemma resealing upon injury, for example, might cause muscular dystrophy [61].

2.2. Membrane resealing mechanism: from the patch hypothesis to acid sphingomyelinase-mediated compensatory endocytosis

The mechanism by which lysosomes reseal damaged plasma membranes was first proposed by Reddy and collaborators in 2001 [62]. Using non-professional secretory cells, such as epithelial cells, myoblasts and fibroblasts, they showed that membrane injury upon scratching is able to trigger lysosomal exocytosis in a Ca^{2+}-regulated manner. Similarly to neuronal synaptic vesicles that have a Ca^{2+}-sensor protein called synaptotagmin I (syt-I) [63], lysosomes have an isoform of synaptotagmin named syt-VII [64, 65]. Synaptotagmins are proteins that have a short ectodomain (N terminus lumenal domain), a transmembrane region and two cytoplasmic domains C2A and C2B that are Ca^{2+}-sensor domains. Reddy and colleagues demonstrated that the C2A domain is the one responsible for regulating Ca^{2+}-dependent lysosomal exocytosis [62]. Since then, it had been shown that lysosomes are able to undergo exocytosis in order to reseal PM injuries generated by different sources, such as pathogens [31] and pore-forming toxins [66], other than mechanical wounding. The most accepted model for PM repair in nucleated cells was proposed in the early 2000s and was called 'The Patch Hypothesis'. According to that model, right underneath the injured site lysosomes underwent chaotic fusion events in which they either fused directly with the PM or with one another in a homotypical fusion manner. Those abnormally enlarged vesicles ended up fusing with the injured PM donating membrane to seal the wounded region [67, 68]. However, the patch model failed to explain the repair caused by pore-forming toxins, which stably binds to the membrane. Later, it was shown that the wounding caused by pore-forming toxins led to the formation of intracellular vesicles.

Wound healing experiments performed in the presence of gold-BSA, added prior to injury, demonstrated that those vesicles have an endocytic origin given that they retained gold-BSA in their lumen [15, 69]. Nonetheless, lysosomes play a pivotal role in the endocytosis-mediated plasma membrane resealing model. Following membrane lesion and increase in intracellular Ca^{2+}, those organelles undergo exocytosis and secrete their enzymes into the extracellular medium. Acid sphingomyelinase (ASM) is one of the enzymes that remain active extracellularly after secretion, generating ceramide as a product of sphingomyelin hydrolysis [70, 71]. Ceramide coalesces at the membrane forming highly ordered domains excluding other lipids, such as glycerophospholipids, from those patches [72]. Those domains induce membrane curvature and budding [71, 73, 74] dragging the injured region inward, in a processes called compensatory endocytosis, closing the wound. Cells either deficient in ASM or pharmacologically inhibited fail to undergo compensatory endocytosis but still trigger lysosomal exocytosis. Addition of recombinant ASM to the extracellular medium is able to restore compensatory endocytosis in those cells [15]. Other lysosomal enzymes are also important to regulate the process. It has been proposed that cysteine proteases, cathepsins B and L, released during lysosomal exocytosis may contribute to facilitate ASM access to PM [75]. Additionally, cathepsin D, another lysosomal enzyme released upon exocytosis, becomes active only later after its release and is responsible for negatively modulating ASM activity, closing the wounding cycle [75]. **Figure 2** depicts a timeline illustrating the evolution of the experimental models that explains how Ca^{2+}-dependent membrane resealing upon lysosomal exocytosis is regulated within cells.

2.3. *Trypanosoma cruzi*: how this parasite can take advantage of intracellular endocytic route to perpetuate its intracellular cycle: the essential role of lysosomes in the process

2.3.1. Trypanosoma cruzi and Chagas disease

T. cruzi is an obligatory flagellated intracellular parasite that causes Chagas disease in human hosts. This pathogen was first identified by the Brazilian doctor Carlos Chagas, in 1909, who not only identified the parasite but also unravelled its life cycle, the invertebrate host, the domestic reservoirs and the symptoms of the disease [35]. *T. cruzi* has a complex life cycle that consists of colonizing the midgut of an invertebrate host (a reduviid bug, also known as 'kissing bug') and several tissues from vertebrate hosts [76, 77]. The cycle on the invertebrate host begins when the reduviid bug takes a blood meal from a mammalian host, containing the trypomastigote forms of the parasite. The bloodstream trypomastigotes are the parasite infective form on vertebrate hosts. Once inside the insect midgut, the parasite differentiates into the epimastigote form, capable of replicating in the invertebrate host. Epimastigotes attach to the waxy walls of the insect hindgut where they differentiate into the metacyclic form in a process known as metacyclogenesis [78].

During a blood meal, the insect excretes, together with the urine and faeces, the metacyclic trypomastigotes, which are capable of infecting the vertebrate host. These released trypomastigotes reach the mammalian host bloodstream either via the wound site or through mucous membranes. Once inside the vertebrate host the metacyclic trypomastigotes can infect a plethora of nucleated cells. When the parasite invades the host cell, it can differentiate into the amastigote form, which is the replicative form on the mammalian host. After several rounds of replication, the amastigotes differentiate into the trypomastigote form and the cells, crowded with parasites, burst open. Extracellular trypomastigotes are now free to perpetuate their cycle and infect new cells and tissues. The process that comprises from intracellular invasion to intracellular multiplication, and cell rupture takes about 4–5 days [79, 80].

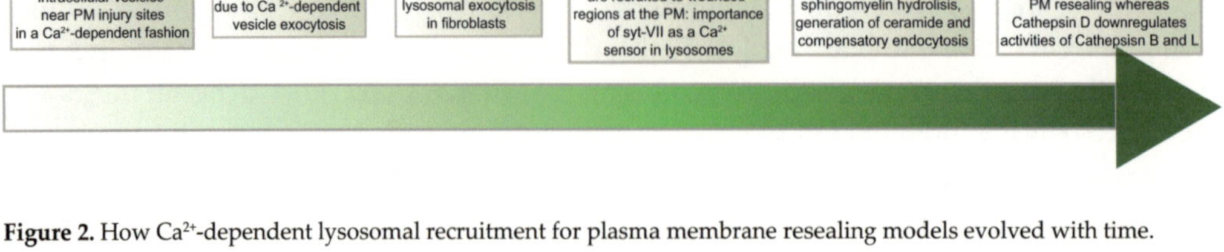

Figure 2. How Ca^{2+}-dependent lysosomal recruitment for plasma membrane resealing models evolved with time.

Recent statistics provided by World Health Organization (WHO) website shows that about 6–7 million people are estimated to be infected with *T. cruzi*, mostly in Latin America [37]. Chagas disease can be transmitted via several routes, being the vectorial route (through the contaminated insect) the canonical one. Along with the vectorial transmission, there are other primary routes of infection, being oral, placental and blood transfusion, the main ones, especially in non-endemic countries where the vector is not present. Those routes are responsible for the worldwide dissemination of Chagas disease [81, 82]. There are other less common ways of acquiring Chagas disease, such as laboratory accidents, dealing with infected animals and organ donation from deceased patients who had Chagas disease [83].

Chagas disease has two phases: acute and chronic. The acute phase lasts from 4 to 8 weeks, and it is usually asymptomatic. However, mild symptoms like fever, for example, might happen 1–2 weeks after infection from the insect vector bite or a month later in other cases of transmission. Only 5–10% of the symptomatic cases might lead to death [84]. The patients who survive from the acute phase will enter chronic phase, which lasts for the patient's lifespan. The majority of the individuals that enter the chronic phase have the indeterminate form of the disease. However, 30–40% of the patients will potentially develop cardiomyopathy, 10% will develop megaesophagus, megacolon, cardiodigestive or neurological problems [37, 85, 86]. Until this day, there is no vaccine available or 100% effective cure for Chagas disease, especially if the disease is diagnosed during the chronic phase. There are two drugs, benznidazole and nifurtimox, which have proven to be effective for some cases during the acute phase. However, their use is limited due to low availability and severe side effects [87].

2.3.2. *Trypanosoma cruzi entry in host mammalian cells*

As mentioned, *T. cruzi* is able to infect most nucleated cells, ranging from professional to non-professional phagocytic cells, the latter being the main focus of the parasite. In order to colonize host cells, *T. cruzi* has to go through four main steps: cell contact and/or attachment to the host cell membrane, intracellular signalling, internalization and intracellular multiplication. Therefore, the events that happen at the plasma membrane are paramount in order to guarantee a successful infection. In order to attach to the host PM, *T. cruzi* uses a variety of proteins that trigger intracellular signalling and parasite entry. Metacyclic trypomastigotes and cell-released trypomastigotes have different repertoires of redundant glycoproteins that have the ability to bind to the extracellular matrix or to specific receptors at the host PM helping in the parasite internalization process (for excellent reviews, please read [38, 88–90]).

Once the parasite gets in contact with host cells, the internalization odyssey takes place. Among them, Ca^{2+} signalling as well as lysosomal recruitment and fusion with the parasitophorous vacuole have been shown to be pivotal for a successful invasion [39, 40] Those two components are also fundamental for modulating PM repair in nucleated mammalian cells, as already described in Section 2.2. We are going to explore on the next subsection, how *T. cruzi* subverts this strategy, used for cells to mend their torn membranes, in order to invade host cells.

2.3.3. Trypanosoma cruzi and lysosomes: importance during parasite entry, maturation and intracellular multiplication

The first evidence showing that *T. cruzi* relies on lysosomes for cell entry was published in 1992 by Tardieux and collaborators [91]. They were motivated by previous work that suggested that *T. cruzi* entry mechanism differed from other pathogens, since actin disruption did not prevent cell invasion in non-professional phagocytic cells [92]. Therefore, *T. cruzi* host cell internalization process was distinct from a phagocytosis-mediated event. If the actomyosin cytoskeleton was not providing the force to drive the parasite towards the host cell cytoplasm, which component of the host cell would be playing this role? It had been shown that when *T cruzi* invades cells it resides temporarily in an acidic vacuole from lysosomal origin [93]. Thus, they decided to investigate how lysosomes participate in that process. They identified lysosomal accumulation near the parasite attachment site at host cell plasma membrane, during the first steps of parasite invasion. By doing perturbations in the microtubule cytoskeleton, in which lysosomes migrate on, they verified changes in *T. cruzi* infection rates. Drug treatments that promoted an outward motion of lysosomes (from the perinuclear region to near the PM) enhanced invasion, whereas blockage of lysosome migration towards the PM inhibited *T. cruzi* entry. Loading lysosomes with sucrose also decreased invasion rate. These authors also showed that cytochalasin D-mediated actin depolymerisation increased invasion by changing lysosomal distribution within the host cell [91]. This work was fundamental for the field since it demonstrated that lysosomes were important for the first stages of parasite invasion, donating membrane for the formation of the parasitophorous vacuole. However, they did not know back then what triggered lysosome secretion during *T. cruzi* invasion.

Two years later, Tardieux and colleagues demonstrated that by exposing NRK cells either to trypomastigotes or to membranes isolated from trypomastigotes, Ca^{2+} transients were elicited in the host cell cytoplasm after only 200 s of exposure, which is faster than the invasion process *per se*, which lasts about 10 min [94]. The same experiment using epimastigotes or epimastigote-isolated membranes, the non-infective form, did not lead to host cell Ca^{2+} transients. Interestingly, they also challenged the cells with *Trypanosoma brucei*, the causative agent of African sleeping sickness, and did not see Ca^{2+} response either. In addition to that, they demonstrated that by blocking intracellular Ca^{2+} transients, before *T. cruzi* exposure, invasion rates decreased. Treatment of host cells for 4 h with Pertussis toxin, known to uncouple $G_{\alpha i}$ and $G_{\alpha 0}$ from their receptors, impairing intracellular signalling cascade [95], halted intracellular Ca^{2+} transients generated upon *T. cruzi* stimulation, suggesting that the parasite-induced Ca^{2+} signalling was likely linked to phospholipase C (PLC) activation and IP_3-mediated Ca^{2+} release from intracellular stores [96, 97]. These two works were primordial since they linked Ca^{2+}-mediated signalling, evoked by *T. cruzi*, and host cell membrane interaction, to lysosomal recruitment and parasite invasion.

In 1995, two other papers from Dr. Norma Andrews' group demonstrated that a Trypomastigote soluble peptidase (also referred to as Proteolytically Generated Trypomastigote Factor—PGTF) was able to generate Ca^{2+} transients in NRK cells [98]. They also proved that PGTF is an agonist of PLC/IP_3 generating Ca^{2+} transients, ultimately leading to actin cytoskeleton remodelling which facilitates *T. cruzi* invasion [99].

http://dx.doi.org/10.5772/intechopen.69305

Years later, in 2001, Wilkowski and collaborators showed that incubation of phagocytic and non-professional phagocytic cells with phosphatidylinositol 3-kinase (PI3K) inhibitors, prior to *T. cruzi* exposure, reduced the invasion rates in those cells [100]. In addition to that, they also demonstrated that *T. cruzi* trypomastigotes or purified trypomastigote membranes elicited high activation of PI3K and PKB/Akt (protein kinase B) on host cells, which was not detected when cells were incubated with epimastigotes or their isolated membranes [100].

Two years later, Woolsey and collaborators demonstrated that even though lysosomes were important for *T. cruzi* invasion, there was a population of parasites that entered cells via a tight parasitophourous vacuole that were devoid of lysosomal markers and formed exclusively by host cell PM markers [101]. These data indicated that *T. cruzi* would also be able to enter host cells via PM invagination and only later fuse with lysosomes. This invasion process was also shown to be independent of host cell actin and to involve PI3K activation [101].

In 2004, Andrade and Andrews demonstrated that parasites that entered the host cell via PM-invagination mechanism gradually escape cells if they do not associate with lysosomal markers, demonstrating that association with lysosomes was pivotal for a successful invasion [39]. Therefore, in the early 2000s, there were two convergent accepted models for *T. cruzi* cell invasion: one that was mediated by host cell PM invagination with later association with lysosomal markers, and another one that was dependent on early lysosomal association, in which lysosomes fused with the PM donating membrane for parasitophorous vacuole formation.

The fact that *T. cruzi* entry was dependent on Ca^{2+} signalling and lysosomal exocytosis, similarly to the lysosomal-mediated plasma membrane repair (explored in Section 2.2) [15], inspired Fernandes and colleagues to investigate whether the parasite would subvert this process to gain access to the host cell. First, they demonstrated that extracellular Ca^{2+} chelation inhibited *T. cruzi* invasion significantly, showing that intracellular Ca^{2+} stores were not the only source during parasite invasion. They also showed that *T. cruzi* causes host cell PM injuries that are rapidly sealed in the presence of Ca^{2+} [40]. Additionally, the concomitant incubation of streptolysin O (SLO), a pore-forming toxin that binds to cholesterol-enriched domains at the PM [102], and *T. cruzi* increased invasion rate, reinforcing the role of extracellular Ca^{2+} in parasite entry process [40]. Finally, pharmacological inhibition or siRNA for ASM (the enzyme responsible for compensatory endocytosis and membrane resealing) reduced trypomastigote invasion, while addition of purified ASM to the extracellular media in ASM depleted cell cultures restored *T. cruzi* invasion. The latter strongly suggested that *T. cruzi* depends on compensatory endocytosis for entering host cells. The proof that compensatory endocytosis was in fact the route of invasion came from their findings showing *T. cruzi* parasitophorous vacuole decorated with ceramide markers. As we already mentioned in Section 2.2, ceramide is generated by ASM-mediated cleavage of sphingomyelin, and it is responsible for the endocytic-directed events following lysosomal fusion. This ceramide containing vacuole was shown to fuse later with lysosomes, providing the anchoring force to retain the parasites inside the host cell [40]. This work set in stone the fact that *T. cruzi* subverts the physiological process by which lysosomes fuse with the PM upon injury in order to successfully invade cells.

Lysosome fusion with plasma membrane induced upon membrane injury is a tightly regulated process and dependent on PM cholesterol content [103]. In 2012, Hissa and collaborators demonstrated that cholesterol depletion of cardiomyocytes prior to exposure to trypomastigotes changed the distribution of lysosomes within the host cell and evoked a massive lysosomal exocytosis near the cell cortex, even in the absence of extracellular Ca^{2+} [104]. These critical lysosomal exocytic events led to a decrease in parasite internalization and lysosomal association for parasitophorous vacuole maturation [104]. One year later, Hissa and colleagues proposed a mechanism by which cholesterol depletion triggered intracellular Ca^{2+}-independent lysosomal secretion. Using methyl-beta cyclodextrin (MβCD) to chelate cholesterol from PM, they showed, by measuring mechanical properties of cell cortices, that cholesterol-depleted cells become more rigid with less membrane fluctuations [105]. This work corroborated previous studies done in cholesterol-depleted endothelial cells [106]. In line with that, cholesterol depletion induced Rho activation, which in turn led to actin polymerization enhancing cortical rigidity. Most importantly, the authors showed that lysosomal exocytosis triggered upon cholesterol depletion was not only Ca^{2+} but also Syt-VII independent, pointing out to a non-regulated secretion of those organelles. They suggested that actin polymerization induced by cholesterol depletion was responsible for the secretion of a lysosomal pool near the cell cortex. Based on these results, one can conclude that cells should have at least two different pools of these organelles, one located closer to the cell cortex, and most likely to be involved with membrane resealing events, and the second located closer to the cell nuclei and probably related to intracellular digestion. For the first pool, actin polymerization could work as an exocytic driving force, whereas for the second, it would present as a barrier for fusion with the PM. In fact, treatment of cells with Latrunculin-A, an actin filament-disrupting drug, induced the secretion of a more internally localized lysosomal pool [105]. In 2015, Hissa and Andrade demonstrated that *T. cruzi* preferentially uses cortical, cholesterol depletion-sensitive lysosomal pool as opposed to the more internally localized, Latrunculin-A sensitive lysosomal reservoir, linking the cortical pool of lysosomes with plasma membrane repair [107].

Regarding intracellular development, *T. cruzi* association with lysosomes remains crucial. In order to replicate in the host cytosol, trypomastigotes need to escape from the lysosomal-enriched parasitophorous vacuole and differentiate into amastigote form. In the late 80s and the early 90s, it was shown that *T. cruzi* secretes a hemolysin factor, active in low pH (5.5), which was capable of lysing erythrocytes isolated from different animal species by forming a large pore in their membranes [108]. If the acidic nature of the parasitophorous vacuole was altered, by raising its pH, parasites were unable to escape to the cytosol. These data corroborated the existence of a hemolysin protein secreted by the parasite, identified as Tc-Tox, which would form a pore at the vacuolar membrane allowing *T. cruzi* to exit and fall into the host cell cytosol [109].

Lysosomal membrane proteins are also important for *T. cruzi* entry and intracellular development, multiplication and release. By using LAMP1/2 knockout cells (LAMP-1/2 KO), Albertti and collaborators showed that LAMP-1 and 2 were important for parasite invasion. Absence of LAMP led to a decrease in parasite ability to invade host cells. Moreover, they showed that, even though parasite entry was reduced, intracellular multiplication was faster in those LAMP-1/2 KO cells, and more trypomastigotes were released after 96 h of infection [110].

http://dx.doi.org/10.5772/intechopen.69305

Those results point out to the importance of these highly sialilated lysosomal proteins for parasite invasion and intracellular development.

As exposed here, opposite to other pathogens, *T. cruzi* takes advantage of lysosomes to infect and perpetuate its life cycle in the vertebrate hosts. It hijacks lysosomes and the physiological route that cells use to repair their torn plasma membranes in order to successfully invade them. Later, it uses lysosomal membrane and acidic environment to gain access to host cell cytosol and colonize it. Any perturbations that prevent lysosomal association with the parasitophorous vacuole or lysosomal distribution, such as host cell PM cholesterol content, culminate with parasite escape and consequently less invasion. Besides, lysosomal content and membrane alterations may also interfere with parasite intracellular development. Therefore, host cell lysosomes control and/or interfere with parasite entry, development, and extracellular release.

Acknowledgements

We would like to acknowledge the following Brazilian funding agencies: Coordenação de Aperfeiçoamento de Pessoal de Nível Superior (CAPES), Conselho Nacional de Desenvolvimento Científico e Tecnológico (CNPq), Fundação de Amparo à Pesquisa de Minas Gerais (FAPEMIG) and Instituto Nacional de Ciência de Tecnologia de Fluidos Complexos (INCT-FCx).

Author details

Barbara Hissa[1*] and Luciana O. Andrade[2]

*Address all correspondence to: barbarahissa@uchicago.edu

1 James Franck Institute, Institute for Biophysical Dynamics and Physics Department, The University of Chicago, Chicago, IL, USA

2 Morphology Department, Biological Sciences Institute, Federal University of Minas Gerais, Belo Horizonte, MG, Brazil

References

[1] Duve CD, Pressman BC, Gianetto R, Wattiaux R, Appelmans F. Tissue fractionation studies 6. Intracellular distribution patterns of enzymes in Rat-Liver tissue. Biochemical Journal. 1955;**60**(1-4):604-617

[2] Hers HG, Berthet J, Berthet L, De Duve C. The hexose-phosphatase system. III. Intracellular localization of enzymes by fractional centrifugation. Bulletin De La Societe De Chimie Biologique Journal (Paris). 1951;**33**(1-2):21-41

[3] Berthet J, Berthet L, Appelmans F, Deduve C. Tissue fractionation studies 2. The nature of the linkage between acid phosphatase and mitochondria in Rat-Liver tissue. Biochemical Journal. 1951;**50**(2):182-189

[4] Novikoff AB, Beaufay H, De Duve C. Electron microscopy of lysosomerich fractions from rat liver. Journal of Biophysical and Biochemical Cytology. 1956;**2**(4 Suppl):179-184

[5] Li XR, Rydzewski N, Hider A, Zhang XL, Yang JS, Wang WY, et al. A molecular mechanism to regulate lysosome motility for lysosome positioning and tubulation. Nature Cell Biology. 2016;**18**(4):404

[6] Dean RT. Direct evidence of importance of lysosomes in degradation of intracellular proteins. Nature. 1975;**257**(5525):414-416

[7] Luzio JP, Pryor PR, Bright NA. Lysosomes: Fusion and function. Nature Reviews Molecular Cell Biology. 2007;**8**(8):622-632

[8] Eskelinen EL, Saftig P. Autophagy: A lysosomal degradation pathway with a central role in health and disease. Biochimica Et Biophysica Acta-Molecular Cell Research. 2009; **1793**(4):664-673

[9] Desjardins M. Biogenesis of phagolysosomes—the kiss and run hypothesis. Trends in Cell Biology. 1995;**5**(5):183-186

[10] Bright NA, Davis LJ, Luzio JP. Endolysosomes are the principal intracellular sites of acid hydrolase activity. Current Biology. 2016;**26**(17):2233-2245

[11] Levine B, Klionsky DJ. Development by self-digestion: Molecular mechanisms and biological functions of autophagy. Developmental Cell. 2004;**6**(4):463-477

[12] Lacombe J, Karsenty G, Ferron M. Regulation of lysosome biogenesis and functions in osteoclasts. Cell Cycle. 2013;**12**(17):2744-2752

[13] Saric A, Hipolito VEB, Kay JG, Canton J, Antonescu CN, Botelho RJ. mTOR controls lysosome tubulation and antigen presentation in macrophages and dendritic cells. Molecular Biology of the Cell. 2016;**27**(2):321-333

[14] Soccio RE, Breslow JL. Intracellular cholesterol transport. Arteriosclerosis Thrombosis and Vascular Biology. 2004;**24**(7):1150-1160

[15] Tam C, Idone V, Devlin C, Fernandes MC, Flannery A, He XX, et al. Exocytosis of acid sphingomyelinase by wounded cells promotes endocytosis and plasma membrane repair. Journal of Cell Biology. 2010;**189**(6):1027-1038

[16] Mrschtik M, Ryan KM. Lysosomal proteins in cell death and autophagy. FEBS Journal. 2015;**282**(10):1858-1870

[17] Parkinson-Lawrence EJ, Shandala T, Prodoehl M, Plew R, Borlace GN, Brooks DA. Lysosomal storage disease: Revealing lysosomal function and physiology. Physiology. 2010;**25**(2):102-115

[18] Mellman I. Endocytosis and molecular sorting. Annual Review of Cell and Developmental Biology. 1996;**12**:575-625

[19] Kumari S, Swetha MG, Mayor S. Endocytosis unplugged: Multiple ways to enter the cell. Cell Research. 2010;**20**(3):256-275

[20] Maritzen T, Schachtner H, Legler DF. On the move: Endocytic trafficking in cell migration. Cellular and Molecular Life Sciences. 2015;**72**(11):2119-2134

[21] Gruenberg J, van der Goot FG. Mechanisms of pathogen entry through the endosomal compartments. Nature Reviews Molecular Cell Biology. 2006;**7**(7):495-504

[22] Maurin M, Raoult D. Q fever. Clinical Microbiology Reviews. 1999;**12**(4):518-553

[23] Hackstadt T, Williams JC. Biochemical stratagem for obligate parasitism of eukaryotic cells by *Coxiella burnetii*. Proceedings of the National Academy of Sciences of the United States of America. 1981;**78**(5):3240-3244

[24] Maurin M, Benoliel AM, Bongrand P, Raoult D. Phagolysosomes of *Coxiella burnetii*-infected cell lines maintain an acidic pH during persistent infection. Infection and Immunity. 1992;**60**(12):5013-5016

[25] Omsland A, Cockrell DC, Howe D, Fischer ER, Virtaneva K, Sturdevant DE, et al. Host cell-free growth of the Q fever bacterium *Coxiella burnetii*. Proceedings of the National Academy of Sciences of the United States of America. 2009;**106**(11):4430-4434

[26] Kohler LJ, Roy CR. Biogenesis of the lysosome-derived vacuole containing *Coxiella burnetii*. Microbes and Infection. 2015;**17**(11-12):766-771

[27] Voth DE, Heinzen RA. Lounging in a lysosome: The intracellular lifestyle of *Coxiella burnetii*. Cell Microbiology. 2007;**9**(4):829-840

[28] Chen JW, Murphy TL, Willingham MC, Pastan I, August JT. Identification of two lysosomal membrane glycoproteins. Journal of Cell Biology. 1985;**101**(1):85-95

[29] Schulze-Luehrmann J, Eckart RA, Olke M, Saftig P, Liebler-Tenorio E, Luhrmann A. LAMP proteins account for the maturation delay during the establishment of the *Coxiella burnetii*-containing vacuole. Cell Microbiology. 2016;**18**(2):181-194

[30] Chappuis F, Sundar S, Hailu A, Ghalib H, Rijal S, Peeling RW, et al. Visceral leishmaniasis: What are the needs for diagnosis, treatment and control? Nature Reviews Microbiology. 2007;**5**(11):873-882

[31] Forestier CL, Machu C, Loussert C, Pescher P, Spath GF. Imaging host Cell-Leishmania interaction dynamics implicates parasite motility, lysosome recruitment, and host cell wounding in the infection process. Cell Host & Microbe. 2011;**9**(4):319-330

[32] McGee ZA, Stephens DS, Hoffman LH, Schlech WF, 3rd, Horn RG. Mechanisms of mucosal invasion by pathogenic Neisseria. Reviews of Infectious Diseases. 1983;**5**(Suppl 4): S708-S714

[33] Ayala BP, Vasquez B, Clary S, Tainer JA, Rodland K, So M. The pilus-induced Ca^{2+} flux triggers lysosome exocytosis and increases the amount of Lamp1 accessible to Neisseria IgA1 protease. Cell Microbiology. 2001;**3**(4):265-275

[34] Ayala P, Lin L, Hopper S, Fukuda M, So M. Infection of epithelial cells by pathogenic neisseriae reduces the levels of multiple lysosomal constituents. Infection and Immunity. 1998;**66**(10):5001-5007

[35] Chagas C. Nova tripanozomiaze humana: estudos sobre a morfolojia e o ciclo evolutivo do Schizotrypanum cruzi n. gen., n. sp., ajente etiolojico de nova entidade morbida do homem. Memórias do Instituto Oswaldo Cruz. 1909;**1**(2):159-218

[36] Bonney KM. Chagas disease in the 21st Century: A public health success or an emerging threat? Parasite. 2014;**21**:11-21

[37] Media Centre—Fact sheets—Chagas Disease (American trypanosomiasis) World Health Organization website. 2016 Available from: http://www.who.int/mediacentre/factsheets/fs340/en/

[38] Barrias ES, de Carvalho TMU, De Souza W. *Trypanosoma cruzi*: Entry into mammalian host cells and parasitophorous vacuole formation. Frontiers in Immunology. 2013;**4**:186-196

[39] Andrade LO, Andrews NW. Lysosomal fusion is essential for the retention of *Trypanosoma cruzi* inside host cells. Journal of Experimental Medicine. 2004;**200**(9):1135-1143

[40] Fernandes MC, Cortez M, Flannery AR, Tam C, Mortara RA, Andrews NW. *Trypanosoma cruzi* subverts the sphingomyelinase-mediated plasma membrane repair pathway for cell invasion. Journal of Experimental Medicine. 2011;**208**(5):909-921

[41] LeSage GD, Robertson WE, Baumgart MA. Bile acid-dependent vesicular transport of lysosomal enzymes into bile in the rat. Gastroenterology. 1993;**105**(3):889-900

[42] Carini R, Trincheri NF, Alchera E, De Cesaris MG, Castino R, Splendore R, et al. PI3K-dependent lysosome exocytosis in nitric oxide-preconditioned hepatocytes. Free Radical Biology & Medicine. 2006;**40**(10):1738-1748

[43] Febbraio M, Silverstein RL. Identification and characterization of LAMP-1 as an activation-dependent platelet surface glycoprotein. Journal of Biological Chemistry. 1990; **265**(30):18531-18537

[44] Sodergren AL, Svensson Holm AC, Ramstrom S, Lindstrom EG, Grenegard M, Ollinger K. Thrombin-induced lysosomal exocytosis in human platelets is dependent on secondary activation by ADP and regulated by endothelial-derived substances. Platelets. 2016;**27**(1):86-92

[45] Hirano T, Saluja A, Ramarao P, Lerch MM, Saluja M, Steer ML. Apical secretion of lysosomal enzymes in rabbit pancreas occurs via a secretagogue regulated pathway and is increased after pancreatic duct obstruction. Journal of Clinical Investigation. 1991;**87**(3):865-869

[46] Jin RU, Mills JC. RAB26 coordinates lysosome traffic and mitochondrial localization. Journal of Cell Science. 2014;**127**(Pt 5):1018-1032

[47] Tapper H, Sundler R. Role of lysosomal and cytosolic pH in the regulation of macrophage lysosomal enzyme secretion. Biochemical Journal. 1990;**272**(2):407-414

[48] Becker SM, Delamarre L, Mellman I, Andrews NW. Differential role of the Ca(2+) sensor synaptotagmin VII in macrophages and dendritic cells. Immunobiology. 2009; **214**(7):495-505

[49] Baron R, Neff L, Brown W, Louvard D, Courtoy PJ. Selective internalization of the apical plasma membrane and rapid redistribution of lysosomal enzymes and mannose 6-phosphate receptors during osteoclast inactivation by calcitonin. Journal of Cell Science. 1990;**97**(Pt 3):439-447

[50] Zhao H, Ito Y, Chappel J, Andrews NW, Teitelbaum SL, Ross FP. Synaptotagmin VII regulates bone remodeling by modulating osteoclast and osteoblast secretion. Developmental Cell. 2008;**14**(6):914-925

[51] Borregaard N, Kjeldsen L, Lollike K, Sengelov H. Granules and secretory vesicles of the human neutrophil. Clinical & Experimental Immunology. 1995;**101**(Suppl 1):6-9

[52] Lindmark IM, Karlsson A, Serrander L, Francois P, Lew D, Rasmusson B, et al. Synaptotagmin II could confer Ca(2+) sensitivity to phagocytosis in human neutrophils. Biochimica et Biophysica Acta. 2002;**1590**(1-3):159-166

[53] Miyake K, McNeil PL. Vesicle accumulation and exocytosis at sites of plasma membrane disruption. Journal of Cell Biology. 1995;**131**(6):1737-1745

[54] Coorssen JR, Schmitt H, Almers W. Ca^{2+} triggers massive exocytosis in Chinese hamster ovary cells. European Molecular Biology Organization. 1996;**15**(15):3787-3791

[55] Rodriguez A, Webster P, Ortego J, Andrews NW. Lysosomes behave as Ca^{2+}-regulated exocytic vesicles in fibroblasts and epithelial cells. Journal of Cell Biology. 1997;**137**(1): 93-104

[56] Lin YC, Ho CH, Grinnell F. Fibroblasts contracting collagen matrices form transient plasma membrane passages through which the cells take up fluorescein isothiocyanate-dextran and Ca^{2+}. Molecular Biology of the Cell. 1997;**8**(1):59-71

[57] McNeil PL, Khakee R. Disruptions of muscle-fiber plasma-membranes—role in exercise-induced damage. American Journal of Pathology. 1992;**140**(5):1097-1109

[58] Clarke MS, Caldwell RW, Chiao H, Miyake K, McNeil PL. Contraction-induced cell wounding and release of fibroblast growth factor in heart. Circulation Research. 1995; **76**(6):927-934

[59] McNeil PL, Ito S. Gastrointestinal cell plasma membrane wounding and resealing in vivo. Gastroenterology. 1989;**96**(5 Pt 1):1238-1248

[60] McNeil PL, Steinhardt RA. Loss, restoration, and maintenance of plasma membrane integrity. The Journal of Cell Biology. 1997;**137**(1):1-4

[61] Bansal D, Campbell KP. Dysferlin and the plasma membrane repair in muscular dystrophy. Trends in Cell Biology. 2004;**14**(4):206-213

[62] Reddy A, Caler EV, Andrews NW. Plasma membrane repair is mediated by Ca(2+)-regulated exocytosis of lysosomes. Cell. 2001;**106**(2):157-169

[63] Sudhof TC, Rizo J. Synaptotagmins: C2-domain proteins that regulate membrane traffic. Neuron. 1996;**17**(3):379-388

[64] Martinez I, Chakrabarti S, Hellevik T, Morehead J, Fowler K, Andrews NW. Synaptotagmin VII regulates Ca(2+)-dependent exocytosis of lysosomes in fibroblasts. The Journal of Cell Biology. 2000;**148**(6):1141-1149

[65] Caler EV, Chakrabarti S, Fowler KT, Rao S, Andrews NW. The Exocytosis-regulatory protein synaptotagmin VII mediates cell invasion by *Trypanosoma cruzi*. Journal of Experimental Medicine. 2001;**193**(9):1097-1104

[66] Roy D, Liston DR, Idone VJ, Di A, Nelson DJ, Pujol C, et al. A process for controlling intracellular bacterial infections induced by membrane injury. Science. 2004;**304**(5676):1515-1518

[67] McNeil PL. Repairing a torn cell surface: Make way, lysosomes to the rescue. Journal of Cell Science. 2002;**115**(Pt 5):873-879

[68] McNeil PL, Kirchhausen T. An emergency response team for membrane repair. Nature Reviews Molecular Cell Biology. 2005;**6**(6):499-505

[69] Idone V, Tam C, Goss JW, Toomre D, Pypaert M, Andrews NW. Repair of injured plasma membrane by rapid Ca2+-dependent endocytosis. Journal of Cell Biology. 2008; **180**(5):905-914

[70] Schuchman EH. Acid sphingomyelinase, cell membranes and human disease: Lessons from Niemann-Pick disease. FEBS Letters. 2010;**584**(9):1895-1900

[71] Gulbins E, Kolesnick R. Raft ceramide in molecular medicine. Oncogene. 2003;**22**(45): 7070-7077

[72] Kolesnick RN, Goni FM, Alonso A. Compartmentalization of ceramide signaling: Physical foundations and biological effects. Journal of Cellular Physiology. 2000;**184**(3):285-300

[73] van Blitterswijk WJ, van der Luit AH, Veldman RJ, Verheij M, Borst J. Ceramide: Second messenger or modulator of membrane structure and dynamics? Biochemical Journal. 2003;**369**:199-211

[74] Draeger A, Babiychuk EB. Ceramide in plasma membrane repair. Handbook of Experimental Pharmacology. 2013;**216**:341-353

[75] Castro-Gomes T, Corrotte M, Tam C, Andrews NW. Plasma Membrane repair is regulated extracellularly by proteases released from lysosomes. PLoS One. 2016;**11**(3):e0152583

[76] Brener Z. Life cycle of *Trypanosoma cruzi*. Revista do Instituto de Medicina Tropical de São Paulo. 1971;**13**(3):171-178

[77] Tyler KM, Engman DM. The life cycle of *Trypanosoma cruzi* revisited. International Journal for Parasitology. 2001;**31**(5-6):472-481

[78] Bonaldo MC, Souto-Padron T, de Souza W, Goldenberg S. Cell-substrate adhesion during *Trypanosoma cruzi* differentiation. Journal of Cell Biology. 1988;**106**(4):1349-1358

[79] Burleigh BA, Andrews NW. The mechanisms of *Trypanosoma-Cruzi* invasion of mammalian-cells. Annual Review of Microbiology. 1995;**49**:175-200

[80] Andrade LO, Andrews NW. The *Trypanosoma cruzi*-host-cell interplay: Location, invasion, retention. Nature reviews Microbiology. 2005;**3**(10):819-823

[81] Munoz J, Prat JGI, Gallego M, Gimeno F, Trevino B, Lopez-Chejade P, et al. Clinical profile of *Trypanosoma cruzi* infection in a non-endemic setting: Immigration and Chagas disease in Barcelona (Spain). Acta Tropica. 2009;**111**(1):51-55

[82] Jackson Y, Getaz L, Wolff H, Holst M, Mauris A, Tardin A, et al. Prevalence, clinical staging and risk for Blood-Borne transmission of chagas disease among Latin American Migrants in Geneva, Switzerland. Plos Neglected Tropical Diseases. 2010;**4**(2):e592-e599

[83] Coura JR. The main sceneries of Chagas disease transmission. The vectors, blood and oral transmissions—a comprehensive review. Memórias do Instituto Oswaldo Cruz. 2015;**110**(3):277-282

[84] Rassi Jr. A, Rassi A, Marin-Neto JA. Chagas disease. Lancet. 2010;**375**(9723):1388-1402

[85] Coura JR, Borges-Pereira J. Chagas disease: 100 years after its discovery. A systemic review. Acta Tropica. 2010;**115**(1-2):5-13

[86] Prata A. Clinical and epidemiological aspects of Chagas disease. The Lancet Infectious Diseases. 2001;**1**(2):92-100

[87] Chatelain E. Chagas disease research and development: Is there light at the end of the tunnel? Computational and Structural Biotechnology Journal. 2017;**15**:98-103

[88] Epting CL, Coates BM, Engman DM. Molecular mechanisms of host cell invasion by *Trypanosoma cruzi*. Experimental Parasitology. 2010;**126**(3):283-291

[89] Yoshida N. Molecular basis of mammalian cell invasion by *Trypanosoma cruzi*. Anais Da Academia Brasileira De Ciencias. 2006;**78**(1):87-111

[90] Watanabe Costa R, da Silveira JF, Bahia D. Interactions between *Trypanosoma cruzi* secreted proteins and host cell signaling pathways. Frontiers in Microbiology. 2016;**7**:388

[91] Tardieux I, Webster P, Ravesloot J, Boron W, Lunn JA, Heuser JE, et al. Lysosome recruitment and fusion are early events required for trypanosome invasion of Mammalian-Cells. Cell. 1992;**71**(7):1117-1130

[92] Schenkman S, Robbins ES, Nussenzweig V. Attachment of *Trypanosoma-Cruzi* to Mammalian-Cells requires parasite energy, and invasion can be independent of the Target-Cell cytoskeleton. Infection and Immunity. 1991;**59**(2):645-654

[93] de Meirelles Mde N, de Araujo Jorge TC, de Souza W, Moreira AL, Barbosa HS. *Trypanosoma cruzi*: phagolysosomal fusion after invasion into nonprofessional phagocytic cells. Cell Structure and Function. 1987;**12**(4):387-393

[94] Tardieux I, Nathanson MH, Andrews NW. Role in host cell invasion of *Trypanosoma cruzi*-induced cytosolic-free Ca2+ transients. The Journal of Experimental Medicine. 1994;**179**(3):1017-1022

[95] Mangmool S, Kurose H. G(i/o) protein-dependent and -independent actions of Pertussis Toxin (PTX). Toxins (Basel). 2011;**3**(7):884-899

[96] Hughes AR, Putney Jr JW. Inositol phosphate formation and its relationship to calcium signaling. Environmental Health Perspectives. 1990;**84**:141-147

[97] Thomas AP, Bird GS, Hajnoczky G, Robb-Gaspers LD, Putney Jr JW. Spatial and temporal aspects of cellular calcium signaling. FasebJournal. 1996;**10**(13):1505-1517

[98] Burleigh BA, Andrews NW. A 120-kDa alkaline peptidase from *Trypanosoma cruzi* is involved in the generation of a novel Ca(2+)-signaling factor for mammalian cells. The Journal of Biological Chemistry. 1995;**270**(10):5172-5180

[99] Rodriguez A, Rioult MG, Ora A, Andrews NW. A trypanosome-soluble factor induces IP3 formation, intracellular Ca2+ mobilization and microfilament rearrangement in host cells. Journal of Cell Biology. 1995;**129**(5):1263-1273

[100] Wilkowsky SE, Barbieri MA, Stahl P, Isola EL. *Trypanosoma cruzi*: Phosphatidylinositol 3-kinase and protein kinase B activation is associated with parasite invasion. Experimental Cell Research. 2001;**264**(2):211-218

[101] Woolsey AM, Sunwoo L, Petersen CA, Brachmann SM, Cantley LC, Burleigh BA. Novel PI 3-kinase-dependent mechanisms of trypanosome invasion and vacuole maturation. Journal of Cell Science. 2003;**116**(Pt 17):3611-3622

[102] Bhakdi S, Tranum-Jensen J, Sziegoleit A. Mechanism of membrane damage by streptolysin-O. Infection and Immunity. 1985;**47**(1):52-60

[103] Andrade L. Understanding the role of cholesterol in cellular biomechanics and regulation of vesicular trafficking: The power of imaging. Biomedical Spectroscopy and Imaging [Internet]. 2016;**5**(s1):S101–S117

[104] Hissa B, Duarte JG, Kelles LF, Santos FP, del Puerto HL, Gazzinelli-Guimarães PH, et al. Membrane cholesterol regulates lysosome-plasma membrane fusion events and modulates *Trypanosoma cruzi* invasion of host cells. PLoS Neglected Tropical Diseases. 2012;**6**(3):e1583–e

[105] Hissa B, Pontes B, Roma PMS, Alves AP, Rocha CD, Valverde TM, et al. Membrane cholesterol removal changes mechanical properties of cells and induces secretion of a specific pool of lysosomes. PloS one. 2013;**8**(12):e82988–e

[106] Byfield FJ, Aranda-Espinoza H, Romanenko VG, Rothblat GH, Levitan I. Cholesterol depletion increases membrane stiffness of aortic endothelial cells. Biophysical Journal. 2004;**87**(5):3336-3343

[107] Hissa B, Andrade LD. *Trypasonoma cruzi* uses a specific subset of host cell lysosomes for cell invasion. Parasitology International. 2015;**64**(2):135-138

[108] Andrews NW, Whitlow MB. Secretion by *Trypanosoma cruzi* of a hemolysin active at low pH. Molecular and Biochemical Parasitology. 1989;**33**(3):249-256

[109] Ley V, Robbins ES, Nussenzweig V, Andrews NW. The exit of *Trypanosoma cruzi* from the phagosome is inhibited by raising the pH of acidic compartments. Journal of Experimental Medicine. 1990;**171**(2):401-413

[110] Albertti LaG, Macedo AM, Chiari E, Andrews NW, Andrade LO. Role of host lysosomal associated membrane protein (LAMP) in *Trypanosoma cruzi* invasion and intracellular development. Microbes and Infection. 2010;**12**(10):784-789

Chapter 4

Autophagy-Lysosome Dysfunction in Amyotrophic Lateral Sclerosis and Frontotemporal Lobar Degeneration

Peter M. Sullivan, Xiaolai Zhou and Fenghua Hu

Additional information is available at the end of the chapter

http://dx.doi.org/10.5772/intechopen.69371

Abstract

Amyotrophic lateral sclerosis (ALS) and frontotemporal lobar degeneration (FTLD) are two devastating neurodegenerative diseases. Several lines of evidence suggest that these diseases are part of a continuum with common genetic factors. As researchers uncover more genes associated with ALS/FTLD, studies have shown that majority of these genes regulate lysosome-related processes. Lysosomes play important roles in clearing damaged organelles and proteins through the autophagy-lysosome pathway and clearing extracellular debris by the endolysosomal pathway. Disruption of both the autophagy and endolysosomal pathways has been implicated in ALS/FTLD pathogenesis.

Keywords: autophagy, lysosome, amyotrophic lateral sclerosis (ALS), frontotemporal lobar degeneration (FTLD), neurodegeneration, progranulin (PGRN), TMEM106B, C9orf72, OPTN, p62, TBK1, ubiquilin2 (UBQLN2), TDP-43, FUS, tau, VCP, CHMP2B

1. Introduction

Proper degradation machinery is necessary for neuronal survival, and disruption of lysosomal function is sufficient to cause neurodegeneration [1–4]. To recycle cellular material, cells use two major pathways: autophagy for damaged organelles and long-lived proteins and the ubiquitin-proteasome system (UPS) for short-lived proteins [5, 6]. Autophagy consists of three pathways and each of them ultimately delivers cellular contents to the lysosome for degradation. The pathways are chaperone-mediated autophagy (CMA), which uses HSC70 to recognize specific misfolded proteins; microautophagy, which directly invaginates material into the lysosome; and macroautophagy, which is responsible for the degradation of organelles, protein aggregates, and large protein complexes. Macroautophagy (hereafter referred to as autophagy) is the most common pathway. The autophagy pathways and molecular mechanisms have been recently reviewed elsewhere [7, 8]. The presence of protein aggregates in most neurodegenerative diseases suggests common underlying problem in protein degradation systems. Here, we summarize the connection between the autophagy-lysosome pathway and two neurodegenerative diseases, amyotrophic lateral sclerosis (ALS) and frontotemporal lobar degeneration (FTLD) [9].

ALS is characterized by the loss of upper and lower motor neurons resulting in progressive weakness and ultimately paralysis. Patients survive a median of 3–5 years from disease onset [10]. FTLD is characterized by the degeneration of neurons in the frontal cortex and anterior temporal lobes. This degeneration leads to changes in behavior and language impairment. The subtypes of FTLD can be distinguished by the prominent symptoms, which reflect the area affected by neuron loss [11, 12]. The subtypes are behavioral variant frontotemporal dementia (bvFTD), semantic dementia (SD), and primary nonfluent aphasia (PNFA). Behavioral variant frontotemporal dementia, the most common subtype, is characterized by changes in behavior such as disinhibition, loss of empathy, impaired social skills, and decline in personality. SD is characterized by impaired language comprehension, and PNFA disrupts speech production [9]. These subtypes often overlap and can additionally include Parkinson's disease-like symptoms. Patients survive for a median of 7–11 years after diagnosis. There are no treatments for FTLD [9]. ALS and FTLD symptoms are often present in the same patient with an indication that these diseases have shared etiology [13, 14].

Each disease is also subdivided by molecular pathology depending on the primary components of inclusion bodies, such as Tau, TDP-43 (TAR DNA-Binding Protein 43), FUS (fused in sarcoma), SOD1 (superoxide dismutase 1) and C9 or f72 dipeptide repeats (DPRs) [9, 15]. In 2006, both ALS and FTLD were found to have neuronal inclusions composed largely of TDP-43, an RNA-binding protein, that are also ubiquitin and p62-positive, suggesting that these aggregates were tagged for degradation [16–18]. Additionally, genetic mutations that can lead to the development of both ALS and FTLD have since been discovered. Thus, these two diseases are linked by clinical concurrence, molecular pathology, and genetic overlap [13, 14, 19].

As many new genes have been identified for FTLD and ALS in the last decade, studies have revealed a common theme of these genes functioning in the lysosomal network (**Figure 1**). Some mutations, such as *GRN*, *TMEM106B*, *CHMP2B*, and valosin-containing protein (*VCP*) are associated with disrupted lysosomes and multivesicular bodies (MVB). Other mutations, such as in *p62/SQSTM1*, *OPTN*, ubiquilin2 (*UBQLN2*), and TANK-binding kinase (*TBK1*) directly disrupt selective autophagy and therefore prevent cargo from being degraded. The rest of the mutations have a more complex relationship with autophagy and lysosome function, such as mutations in the RNA-binding proteins TDP-43 and *FUS*. Here, we will discuss the genetic causes of ALS and FTLD in more detail with specific emphasis on lysosomal and autophagy impairment (**Figure 1**).

http://dx.doi.org/10.5772/intechopen.69371

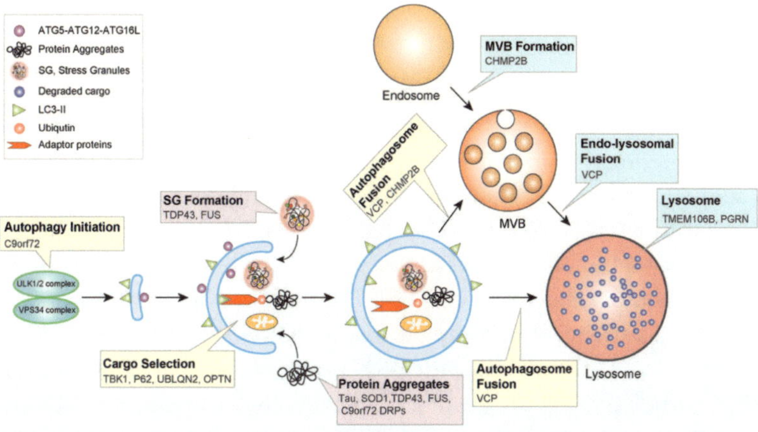

Figure 1. Functions of the ALS/FTLD genes in the autophagy-lysosome pathway. Many genes associated with ALS/FTLD play critical roles in the endosome-lysosomal pathway, regulate lysosomal functions, or affect autophagy pathway directly or indirectly.

2. Mutations affecting the endolysosome pathway: progranulin (PGRN), TMEM106B, CHMP2B, and VCP

2.1. Progranulin

The most common cause of familial FTLD with ubiquitin-positive aggregates is mutation of the *GRN* gene, which accounts for 10% of all FTLD cases and ~25% of familial FTLD [20–22]. About 70 mutations in the *GRN* gene have been linked to FTLD, most of which have been shown or predicted to decrease PGRN protein level or disrupt secretion of PGRN [20–24]. While FTLD is caused by haploinsufficiency of PGRN, a more serve neurodegeneration is caused by homozygous loss of PGRN. This complete loss of PGRN results in neuronal ceroid lipofuscinosis (NCL), a type of lysosome storage disorder (LSD) characterized by the build-up of autofluorescent lipofuscin [25, 26]. These findings suggest that loss of function mutations in the *GRN* gene causes neurodegenerative diseases in a dose-dependent manner and PGRN is important for lysosome function.

The function of PGRN is still under investigation: it is known to be a secreted glycoprotein comprised of 7.5 granulin repeats with pleiotropic roles, including protein homeostasis, inflammation, and neuronal survival and outgrowth [27]. Recently, several lines of evidence suggest that it plays a vital role in lysosome function. First, *GRN* has been found to be regulated with other lysosomal genes [28]. Furthermore, *GRN* mRNA and PGRN protein levels are upregulated in response to lysosome or autophagy inhibition [29]. Finally, PGRN was found to be delivered to the lysosome [30, 31]. PGRN reaches the lysosome through at two independent pathways. In one pathway, PGRN's extreme C-terminus binds the sorting receptor sortilin, which carries PGRN to the lysosome [30, 32]. In the second pathway, PGRN binds prosaposin, and they are transported to the lysosome together by the cation-independent mannose-6-phosphate receptor (CI-M6PR) and low-density lipoprotein receptor-related protein 1 (LRP1) [31].

Mouse models of PGRN deficiency have consistently found increased levels of ubiquitin and p62, an adaptor for delivering cargo to the autophagosome [33], buildup of lipofuscin and its protein components saposin D and SCMAS and electron-dense storage granules, all of which suggest lysosome impairment [34–36]. Several models also found aggregation of TDP-43, similar to what is seen in FTLD patients [34, 37, 38]. Furthermore, PGRN-deficient mouse models also phenocopy FTLD symptoms such as decreased social interaction and mild learning/memory defects [35, 38–40]. The presence of clear lysosomal problems in mouse models and in patients with complete loss of PGRN suggests that PGRN is necessary for lysosome function. FTLD patients with GRN mutations also exhibit typical pathological features of NCL pathology [36], suggesting FTLD and NCL caused by PGRN mutations are pathologically linked and lysosomal dysfunction is one of the underlying disease mechanisms for FTLD-GRN. However, how PGRN regulates lysosomal function remains to be investigated.

2.2. TMEM106B

Another gene associated with FTLD is *TMEM106B*, which is the only identified risk factor for FTLD with *GRN* mutations [41–44]. *TMEM106B* was also found to increase risk in patients with *C9orf72* hexanucleotide repeat expansions [45, 46]. The *TMEM106B* SNP associated with FTLD increases the mRNA and protein levels of TMEM106B [36, 44, 47]. TMEM106B is a single pass, type II transmembrane protein that localizes to the late endosome and lysosome [47–49]. Cellular studies on TMEM106B have pointed to roles in lysosome trafficking and lysosomal stress response [50, 51]. Overexpression of TMEM106B in cells disrupts lysosome morphology and function [47, 48]. Furthermore, when a transgenic TMEM106B mouse line was crossed with a PGRN deficient mouse line, the lysosome abnormalities and lipofuscin accumulation seen in PGRN deficient mice were exacerbated [52]. The connection between TMEM106B's role at the lysosome and a risk factor for FTLD with *GRN* mutations further highlights the importance of the lysosome pathway in FTLD etiology.

2.3. CHMP2B

The sole mutation identified to cause FTLD with ubiquitin-positive aggregates, but tau, TDP-43, and FUS negative inclusions, occurs in the gene *CHMP2B* [53, 54]. *CHMP2B* has also been found to cause rare cases of ALS [55]. CHMP2B functions in the ESCRT-III complex, involved in MVB formation to deliver cargo from endocytic pathway to lysosomes [56, 57]. The mutations identified create an early termination of the protein, resulting in an unregulated CHMP2B truncation that is unable to recruit VPS4 to recycle the ESCRT-III complex to new sites of MVB formation [58, 59]. With ESCRT-III still engaged on the MVB, MVB-lysosome fusion cannot take place [54, 60–62]. Furthermore, *CHMP2B* mutations impair autophagosome maturation, possibly through the disruption of amphisome formation between autophagosome and late endosomes [63–66]. Mouse models of *CHMP2B* mutations replicate both ALS and FTLD pathology, whereas *CHMP2B* knockout mice do not show neurodegenerative phenotypes, implicating a gain of function disease mechanism [67–70]. Similar to the PGRN deficiency mouse models, *CHMP2B* mutations cause protein inclusions and accumulation of autofluorescent aggregates in the frontal cortex, reminiscent of lysosome storage disorders [71]. Thus, FTLD-associated mutations in *CHMP2B* impair the endolysosomal pathway, which may cause additional defects in autophagy [66, 69], providing additional evidence that disruption of the autophagy-lysosome pathway may drive ALS and FTLD.

2.4. VCP

Valosin-containing protein (VCP) has been implicated in several diseases including FTLD [22, 72–76], ALS [77], and Charcot Marie Tooth disease, a genetic peripheral nerve disorder [78]. VCP is an AAA^{+}-ATPase that delivers and unfolds ubiquitinated proteins, as well as endoplasmic reticulum–associated protein degradation (ERAD) substrates, at the proteasome [79–83]. Furthermore, VCP binds to clathrin and EEA1 to regulate the size and selectivity of endosomes [83–85]. Pharmacological inactivation of VCP as well as VCP knockdown inhibits MVB formation and blocks autophagosome maturation, resulting in accumulated LC3-II, ubiquitin, and p62 levels along with cytoplasmic TDP-43 aggregation [86–88]. Disease-associated mutants of VCP present similar phenotypes in transgenic mouse models, whereas complete loss of VCP is embryonic lethal [86, 89–91]. Finally, VCP mutants inhibit the autophagic turnover of stress granules, which may be relevant to the accumulation of TDP-43-positive aggregates found in patients with VCP mutations [76, 92, 93]. The precise mechanism that halts autophagosome maturation in VCP mutations remains unclear, though MVB dysfunction may play a role [66]. VCP's role in MVB formation and autophagic flux suggest that loss of VCP function may cause ALS, FTLD, and other related neurodegenerative diseases by impairing the autophagy-lysosome pathway.

3. Autophagy adaptor proteins

Further evidence that ALS and FTLD are linked to autophagy and lysosome disruption comes from mutations that directly affect several autophagy adaptor proteins and their regulation. Genetic mutations in the adaptor proteins *p62/SQSTM1*, *UBQLN2*, and *OPTN* have been shown to contribute to rare cases of ALS [94–99] and FTLD [100, 101]. All these adaptor proteins contain an ubiquitin-associated (UBA) domain, which is able to bind polyubiquitin conjugated proteins that are tagged for degradation by either the UPS or autophagy. The autophagy adaptors then associate with LC3 on the autophagosome to deliver the cargo for degradation through autophagy-lysosome pathway.

3.1. p62/SQSTM1

p62/SQSTM1 (p62)-positive inclusions have been observed in patient tissue samples in both ALS and FTLD [18, 102–104]. The association of p62 with inclusions suggests that the inclusion body has been targeted for degradation and the accumulation of such inclusions suggests defects with their turnover [33, 105, 106]. p62 bridges autophagy substrates to the autophagosome by interacting with ubiquitinated proteins via its UBA domain [107] and LC3 with its LC3-interacting region (LIR) [33, 108, 109].

p62 is activated by phosphorylation at Ser407 by ULK1, allowing further phosphorylation by casein kinase 2 or TANK-binding kinase 1 (TBK1), which increases p62's affinity for polyubiquitinated cargo [110–113]. p62 acts within the selective autophagy system by aggregating proteins and organelles together for the autophagosome to enclose [106, 114]. These aggregated cargos are then subject to autophagy [115, 116]. While p62 accumulation and association with protein aggregates broadly suggests a defect in autophagy, mutations in p62 directly link selective autophagy impairment to neurodegeneration.

The *p62* mutations identified in ALS and FTLD patients disrupt aggregate formation or decrease the amount of p62 protein produced, leading to loss of function [117–119]. Homozygous mutation of *p62* causes adolescence/childhood-onset neurodegeneration with a defect in mitochondrial depolarization response due to impaired autophagy [120]. Thus, a loss of normal p62 function in autophagy leads to neurodegeneration in a dose-dependent manner, with earlier onset correlating to lower levels of functional p62.

In addition to its role in autophagy, p62 also links ubiquitinated cargo to the proteasome through its UBA domain [106] and mediates the degradation of the protein via the UPS, indicating that p62 plays multiple roles in proteostasis [121].

3.2. Ubiquilin2

Another adaptor protein implicated in ALS and FTLD is ubiquilin2 (UBQLN2) [95, 122]. Similar to p62, UBQLN2 is able to recognize ubiquitinated proteins and bind them via its UBA domain [123]. The UBA domain is also required for UBQLN2 to associate with the autophagosome, though unlike p62 and OPTN, UBQLN2 does not directly recognize LC3 [124, 125].

Knockdown of UBQLN2 in culture reduced autophagosome formation and inhibited lysosomal degradation of mitochondria [124, 125]. This loss of UBQLN2 also sensitizes cells to starvation-induced death in an autophagy-dependent manner [124]. Interestingly, UBQLN2 binds directly to TDP-43 holo-protein and C-terminal fragments and may regulate the levels of TDP-43 in the cell independent of ubiquitin [126]. Indeed, overexpression of UBQLN2 in culture can reduce aggregation of TDP-43 [126].

Many of the disease-associated mutations map to the proline-rich domain in *UBQLN2*, which is important in mediating protein-protein interactions [95, 127]. Furthermore, mutations in *UBQLN2* have a reduced binding to hnRNPA1, a RNA-binding protein associated with stress granules. Interestingly, mutations in hnRNPA1 are also associated with ALS and these mutations also disrupt its interaction with UBQLN2 [128], confirming that the interaction of autophagy adaptors with stress granules is important for neuronal survival.

UBQLN2 knockout in a rodent model showed no neuronal loss, implying that loss of function is not the disease mechanism or that other autophagy adaptors are able to compensate for its loss *in vivo*. Transgenic animals with the ALS/FTLD-associated *UBQLN2* mutations produce ubiquitin, p62, and UBLQN2-positive puncta accompanied by neuronal loss, cognitive defects, and motor impairment [129–131]. Increased expression of the wild-type UBQLN2 also causes neurodegeneration in a rodent model [132]. Thus, unlike mutations in *p62*, *UBQLN2* mutations appear to have a gain of function mechanism that impairs proper protein degradation by autophagy.

In addition to its function in the autophagy pathway, UBLQN2 binds to the proteasome through its ubiquitin-like (UBL) domain to deliver polyubiquitinated proteins and ERAD substrates to the proteasome for degradation [133]. A role of UBQLN2 in delivering protein aggregates to proteasome-mediated degradation via HSP70 has been recently demonstrated [134]. UBQLN2 also function together with other ALS/FTLD-related proteins, such as regulating endosome constitution with OPTN [135] and delivering ERAD substrates to the proteasome with VCP [136].

3.3. Optineurin (OPTN)

Rare mutations in *OPTN* are also associated with both ALS [97, 99]as well as FTLD [101]. These mutations are expected to decrease the level of OPTN protein, suggesting a loss of function resulting in disease [101]. In total, 1–4% of familial ALS cases are linked to mutations in OPTN [137]. OPTN, like p62 and UBQLN2, binds to polyubiquitin-labeled proteins via a UBA domain [138]. OPTN also binds LC3 through an LIR to connect cargo to autophagosomes. Damaged mitochondria specifically recruit OPTN to induce mitophagy [139]. In support of a loss of function model for OPTN, depletion of OPTN in zebrafish causes motor defects [140].

OPTN also interacts with several other proteins associated with ALS. The E3 ubiquitin ligase HACE1 ubiquitinates OPTN to promote binding to p62, which forms a complex that enhances autophagic flux [141]. Similarly, phosphorylation of OPTN by TBK1 increases the interaction of OPTN and p62 to the same effect [138, 142]. OPTN also binds directly to SOD1 aggregates independently of ubiquitination. Mutations in *OPTN* do not affect this interaction, but do impair autophagic clearance of SOD1 protein aggregates through an unknown mechanism [138, 140].

Mutation in *OPTN* had previously been linked to primary open-angle glaucoma (POAG) where these mutations were shown to decrease basal autophagy and inhibit autophagic flux upon autophagy induction [143]. Thus, mutations in *OPTN* have clear links to multiple neurodegenerative disease with consistent impairment in the autophagy pathway. How mutations in the same gene and similar cellular impairments can lead to distinct clinical outcomes remains unclear.

3.4. TBK1

TBK1 has recently been associated with both ALS and FTLD [96, 98, 101, 110, 111, 144–147]. TBK1 has functions in autophagy and in inflammation [148]. Regarding its function in autophagy, TBK1 phosphorylates p62 and OPTN to increase their binding to LC3 and ubiquitin, respectively [138, 142]. Many of the discovered disease-associated mutations are expected to decrease TBK1 protein level, suggesting a loss of function model [96, 101].

While TBK1 interacts with both p62 and OPTN, TBK1 and OPTN share several additional connections. Like OPTN, some mutations in TBK1 also cause glaucoma [149].

Furthermore, the mutation in OPTN that causes POAG enhances the binding of OPTN to TBK1, which may sequester TBK1 and prevent it from carrying out its normal function [142]. Finally, both TBK1 and OPTN are required specifically for mitophagy, with depletion of either component or expression of an ALS-associated mutant impairing mitophagy [150]. Taken together, mutations in *TBK1* cause decreased protein expression and defects in p62 and OPTN regulation again supporting a role of autophagy in preventing ALS and FTLD.

4. C9orf72

The most common known cause of both ALS and FTLD was discovered to be a hexanucleotide intronic repeat expansion in the gene *C9orf72* [151–153]. This repeat expansion is found in 18–25% of familial FTLD, 40% of familial ALS, and 4–8% of sporadic ALS and FTLD combined [154, 155]. While patients with C9orf72 mutations display TDP-43-positive aggregates, they also have separate inclusions unique to this genetic mutation. These ubiquitin, p62, and occasionally UBQLN2-positive inclusions also contain dipeptide repeats generated from the repeat expansion [156–160]. Three molecular mechanisms of disease have been proposed: toxic gain of function of RNA repeats, gain of function of dipeptide repeats (DPRs) produced by repeat-associated non-ATG translation, and haploinsufficiency of the C9orf72 protein.

RNA-repeats transcribed from the repeat expansion form nuclear foci and sequester many RNA-binding proteins, including several RNA-binding proteins already implicated in ALS and FTLD [151, 161–163]. In addition the RNA foci disrupt nucleocytoplasmic transport [164, 165]. Furthermore, five distinct DPRs are translated and can also alter nucleocytoplasmic transport - [167, 168] as well as disrupt membrane-less, phase-separated organelles such as the nucleolus, nuclear pore, and stress granules [169]. Nuclear translocation of TDP-43 has been shown to be blocked by both RNA repeats and DPRs [166-168] , allowing TDP-43 to accumulate and aggregate in the cytosol, which is observed in ALS/FTLD with C9orf72 mutations.

Haploinsufficiency was also proposed as a disease mechanism [153, 151, 170–172]. Early C9orf72-depletion models in *Caenorhabditis elegans* and zebrafish showed motor dysfunction, supporting this model [173, 174]. However, a neuronal-specific C9orf72 knockout mouse showed no such phenotype [175]. Complete C9orf72 knockout mice also do not show much neurodegeneration, but instead exhibit severe immune problems similar to autoimmune disorders [176–181].

Interestingly, C9orf72 has been reported to play a role in autophagy and lysosome regulation. While many of the reports suggest that C9orf72 and its binding partners, SMCR8 and WDR41, play a role in regulating autophagy initiation or maturation, likely via the FIP200/ULK1 complex, the precise mechanism remains uncertain [179, 182–186]. Other reports suggests that C9orf72 plays a role in mammalian Target of Rapamycin (mTOR) and Transcription Factor EB (TFEB) signaling [186, 187], in stress granule assembly [188], or in actin dynamics [189].

5. RNA-binding proteins

The RNA-binding proteins TDP-43 and FUS have been closely associated with ALS and FTLD. Pathogenic TDP-43 or FUS aggregates are present in both conditions, though mutations in these genes result primarily in ALS [190]. Both proteins travel between the nucleus and cytoplasm as they regulate gene splicing, mRNA stability and trafficking, and stress granule dynamics [191, 192].

As both TDP-43 and FUS regulate the RNA from thousands of genes, many cellular problems could be anticipated. However, several lines of evidence have pointed out a role in regulating and challenging the autophagy pathway [193].

5.1. TDP-43

The identification of TDP-43 as the main component of protein aggregates in both ALS and FTLD spurred the awareness that ALS and FTLD had some underlying similarities [16, 17]. Interestingly, mutations in *TARDBP* (TAR DNA binding protein), the gene encoding TDP-43, lead overwhelmingly to ALS or ALS/FTLD, but not to FTLD alone [194, 195]. While soluble TDP-43 can be cleared by chaperone-mediated autophagy through its interaction with Hsc70 [196], TDP-43-positive stress granules and aggregates are cleared by macroautophagy [197, 198].

In addition being a substrate of autophagy, TDP-43 may play a direct role in regulating autophagy through its transcriptional regulation of *ATG7* [199]. As TDP-43 is sequestered in protein aggregates, it can no longer regulate *ATG7* transcription, impairing autophagy initiation, and further promoting TDP-43 accumulation [198, 199]. In a similar manner, TDP-43 also regulates the mRNA for Regulatory-Associated Protein of mTOR (RPTOR) and Dynactin subunit 1 (DCTN1) [197]. *RPTOR* encodes a component of the mTOR complex, and loss of *RPTOR* due to TDP-43 loss of function upregulates lysosome and autophagy biogenesis [197]. However, TDP-43 loss of function also results in reduced *DCTN1* mRNA, which encodes dynactin, a key component of autophagosome-lysosome fusion, leading to the accumulation of autophagosomes, preventing the turnover of aggregated TDP-43 [197].

TDP-43 additionally plays an important role in stress granule dynamics and mutations in *TARDBP* have been shown to increase the stability of stress granules, possibly allowing them to become irreversible protein aggregates [198, 200–203]. In support of this prolonged stress granule hypothesis, mutations in VCP decrease stress granule turnover by autophagy, leading to TDP-43-positive inclusion [92].

The interaction of TDP-43 with autophagy suggests a complex regulatory balance between the two under normal conditions. In disease states, a feedforward mechanism of TDP-43 sequestration into stress granules and aggregates followed by impaired autophagy could drive pathogenesis of ALS and FTLD [9, 202].

5.2. FUS

Like *TARDBP*, mutations in *FUS* have been linked more closely to ALS, though positive protein aggregates for FUS appear in both ALS and FTLD [9]. FUS-positive inclusions account for about 5–10% of FTLD cases [9] and 1% of ALS cases [15]. Several proposed mechanisms link FUS to disruption of the autophagy-lysosome pathway. First, the presence of FUS-positive aggregates in both familial and sporadic cases of ALS and FTLD suggests FUS may be particularly susceptible to aggregation. FUS is also involved in autoregulation, which could allow for a feedforward cycle of increased FUS production followed by cytosolic accumulation and aggregation [198, 204].

Additionally, mutations in *FUS* have been linked to altered stress granule dynamics [205, 206]. FUS-positive stress granules were found to be degraded by autophagy; however, stress granules containing mutant FUS were more stable and prevented stress granules disassembly [198]. As with TDP-43, stabilized stress granules may promote insoluble aggregate formation [202, 207–209]. This increases the burden on the autophagy pathway and may drive further cell damage. A recent study also found that ALS-associated mutant *FUS* was able to inhibit the early steps of autophagosome formation, leading to impaired autophagy flux [210]. Many of these studies found that enhancing autophagy, genetically or pharmaceutically, was able to reduce FUS-positive inclusions and prevent cellular toxicity [198, 205, 210]. While less well understood than TDP-43, the RNA-binding protein FUS seems to play a similar cellular role as TDP-43, including regulating the dynamics of stress granules. Besides increased burden on autophagy due to stabilized stress granules, FUS may also play a more direct role in autophagy impairment.

6. Microtubule-associated protein tau

Thirty percent of familial FTLD cases are caused by mutations in Microtubule-Associated Protein Tau (MAPT), encoding the protein tau [211]. These cases are characterized by the presence of tau aggregates positive for ubiquitin and p62, suggesting impaired degradation of accumulated tau [121, 212]. Genetic disruption of autophagy cargo selection is sufficient to cause aggregation of pathogenic tau [213]. The tau protein is mostly well-known for its association with Alzheimer's disease, when it also forms aggregates and is accompanied by neurodegeneration of the hippocampus [214]. How Alzheimer's disease and FTLD patients have overlapping cellular pathology but develop different clinical symptoms remains unclear.

Full length tau can be degraded by the UPS in an ubiquitin-dependent and independent manner [121, 215, 216], whereas misfolded or phosphorylated tau is sent to the autophagy pathway [217]. Generally, tau aggregation and toxicity correlates with autophagy activity, where enhanced autophagy rescues neurodegeneration and impairment exacerbates the symptoms [218–221]. Likewise, modulating TFEB to increase lysosome biogenesis prevents the accumulation of tau [222].

Tau is a microtubule-binding protein that helps to stabilize axonal microtubules [223, 224]. Small increases in unbound tau induces aggregation, suggesting that even mild impairment

http://dx.doi.org/10.5772/intechopen.69371

of the UPS or autophagy-lysosome pathway could lead to pathological tau accumulation [225, 226]. In support of this idea, Niemann-Pick disease, another lysosome storage disorder, also develops tau aggregates [227, 228]. These studies suggest that tau clearance is highly dependent on autophagy and lysosome function and disruption of this pathway may drive tau aggregation. Furthermore, tau has a role in microtubule stability and disrupted cytoskeletal dynamics and trafficking have also been proposed as a disease mechanism. Since lysosomes, endosomes, MVB, and autophagosomes all move along microtubules, any disruptions would affect their ability to maintain proteostasis [229].

7. Discussion

ALS and FTLD are distinct clinical disorders that share overlapping symptoms, pathology, and genetics. Many of the causative genetic mutations and risk factors result in disruption of the lysosome-autophagy pathway (**Figure 1**). Some disease-associated mutants or alleles directly impact lysosomal function through yet unknown mechanisms, such as *PGRN* and *TMEM106B*, or through disruption of the late stages of the endolysosome pathway, as *VCP* and *CHMP2B* mutations are proposed to do. Beyond the lysosome, there are also many mutations in adaptor proteins that impair selective autophagy, including *p62/SQSTM1, OPTN*, and *UBQLN2*. The misregulation of these adaptors is sufficient to induce neurodegeneration, as seen with *TBK1* mutants. Finally, some mutations have a more intricate relationship to the autophagy-lysosome pathway that future research will have to address, including C9orf72 protein, repeat-associated RNA foci, and dipeptide repeats, as well as the microtubule-binding protein tau and the RNA-binding proteins TDP-43 and FUS.

Identifying the underlying cellular problems that lead to disease is an important step in being able to distinguish disorders and subtypes that may ultimately require distinct diagnosis and treatment. The genetic analysis of ALS and FTLD has improved our understanding of this disease spectrum and may inform us of the broad problems that underlie both familial and sporadic ALS and FTLD. The consistent impairment of cellular clearance pathways by ALS and FTLD-associated mutations points to a disease mechanism that is likely to be shared in undiscovered genetic causes, as well as environmental risk factors, that account for the cases of ALS and FTLD that have no known cause.

Acknowledgements

This work is supported by funding to F.H. from the Alzheimer's Association, the Association of Frontotemporal Dementia (AFTD), and the Muscular Dystrophy Association and NINDS (R21 NS081357-01, R01NS088448-01) and by funding to P.M.S. from the Henry and Samuel Mann Outstanding Graduate Student Award and to X.Z. from the Weill Institute Fleming Postdoctoral fellowship.

Author details

Peter M. Sullivan, Xiaolai Zhou and Fenghua Hu*

*Address all correspondence to: fh87@cornell.edu

Department of Molecular Biology and Genetics, Weill Institute for Cell and Molecular Biology, Cornell University, Ithaca, NY, USA

References

[1] Mizushima N, Hara T. Intracellular quality control by autophagy: How does autophagy prevent neurodegeneration? Autophagy. 2006;**2**(4):302-304

[2] Lee S, Sato Y, Nixon RA. Lysosomal proteolysis inhibition selectively disrupts axonal transport of degradative organelles and causes an Alzheimer's-like axonal dystrophy. The Journal of Neuroscience. 2011;**31**(21):7817-7830

[3] Komatsu M, Waguri S, Chiba T, Murata S, Iwata J, Tanida I, Ueno T, Koike M, Uchiyama Y, Kominami E, Tanaka K. Loss of autophagy in the central nervous system causes neurodegeneration in mice. Nature. 2006;**441**:880-884

[4] Nedelsky NB, Todd PK, Taylor JP. Autophagy and the ubiquitin-proteasome system: Collaborators in neuroprotection. Biochimica et Biophysica Acta Molecular Basis of Disease. 2008;**1782**(12):691-699

[5] Mizushima N. Autophagy : Process and function. Genes & Development. 2007;**21**: 2861-2873

[6] Wong E, Cuervo AM. Integration of clearance mechanisms : The proteasome and autophagy. Cold Spring Harbor Perspectives in Biology. 2010;**2**(12):a006734

[7] Feng Y, Yao Z, Klionsky DJ. How to control self-digestion: Transcriptional, post-transcriptional, and post-translational regulation of autophagy. Trends in Cell Biology. 2015;**25**:354-363

[8] Carroll B, Hewitt G, Korolchuk VI. Autophagy and ageing: Implications for age-related neurodegenerative diseases. Essays in Biochemistry. 2013;**55**:119-131

[9] Gotzl JK, Lang CM, Haass C, Capell A. Impaired protein degradation in FTLD and related disorders. Ageing Research Reviews. 2016;**32**:122-139

[10] Morgan S, Orrell RW. Pathogenesis of amyotrophic lateral sclerosis. British Medical Bulletin. 2016;**119**:87-97

[11] Ratnavalli E, Brayne C, Dawson K, Hodges JR. The prevalence of frontotemporal dementia. Neurology. 2002;**58**(11):1615-1621

[12] Neary D, Snowden JS, Mann DM. Classification and description of frontotemporal dementias. Annals of the New York Academy of Sciences. 2000;**920**:46-51

[13] Hardy J, Rogaeva E. Motor neuron disease and frontotemporal dementia: Sometimes related, sometimes not. Experimental Neurology. 2014;**262**(PB):75-83

[14] Janssens J, Van Broeckhoven C. Pathological mechanisms underlying TDP-43 driven neurodegeneration in FTLD-ALS spectrum disorders. Human Molecular Genetics. 2013; **22**(R1):77-87

[15] Guerrero EN, Wang H, Mitra J, Hegde PM, Stowell SE, Liachko NF, et al. TDP-43/ FUS in motor neuron disease: Complexity and challenges. Progress in Neurobiology. 2016;**145-146**:78-97

[16] Neumann M, Sampathu DM, Kwong LK, Truax AC, Micsenyi MC, Chou TT, et al. Ubiquitinated TDP-43 in frontotemporal lobar degeneration and amyotrophic lateral sclerosis. Science. 2006;**314**(5796):130-133

[17] Arai T, Hasegawa M, Akiyama H, Ikeda K, Nonaka T, Mori H, et al. TDP-43 is a component of ubiquitin-positive tau-negative inclusions in frontotemporal lobar degeneration and amyotrophic lateral sclerosis. Biochemical and Biophysical Research Communications. 2006;**351**(3):602-611

[18] Tanji K, Zhang HX, Mori F, Kakita A, Takahashi H, Wakabayashi K. P62/sequestosome 1 binds to TDP-43 in brains with frontotemporal lobar degeneration with TDP-43 inclusions. Journal of Neuroscience Research. 2012;**90**(10):2034-2042

[19] Ng ASL, Rademakers R, Miller BL. Frontotemporal dementia: A bridge between dementia and neuromuscular disease. Annals of the New York Academy of Sciences. 2015; **1338**(1):71-93

[20] Baker M, Mackenzie IR, Pickering-Brown SM, Gass J, Rademakers R, Lindholm C, et al. Mutations in progranulin cause tau-negative frontotemporal dementia linked to chromosome 17. Nature. 2006;**442**(7105):916-919

[21] Cruts M, Gijselinck I, van der Zee J, Engelborghs S, Wils H, Pirici D, et al. Null mutations in progranulin cause ubiquitin-positive frontotemporal dementia linked to chromosome 17q21. Nature. 2006;**442**(August):920-924

[22] Sieben A, Van Langenhove T, Engelborghs S, Martin JJ, Boon P, Cras P, et al. The genetics and neuropathology of frontotemporal lobar degeneration. Acta Neuropathologica. 2012;**124**(3):353-372

[23] Gass J, Cannon A, Mackenzie IR, Boeve B, Baker M, Adamson J, et al. Mutations in progranulin are a major cause of ubiquitin-positive frontotemporal lobar degeneration. Human Molecular Genetics. 2006;**15**(20):2988-3001

[24] Shankaran SS, Capell A, Hruscha AT, Fellerer K, Neumann M, Schmid B, et al. Missense mutations in the progranulin gene linked to frontotemporal lobar degeneration with ubiquitin-immunoreactive inclusions reduce progranulin production and secretion. The Journal of Biological Chemistry. 2008;**283**(3):1744-1753

[25] Smith KR, Damiano J, Franceschetti S, Carpenter S, Canafoglia L, Morbin M, et al. Strikingly different clinicopathological phenotypes determined by progranulin-mutation dosage. The American Journal of Human Genetics. 2012;**90**(6):1102-1107

[26] Canafoglia L, Morbin M, Scaioli V, Pareyson D, D'Incerti L, Fugnanesi V, et al. Recurrent generalized seizures, visual loss, and palinopsia as phenotypic features of neuronal ceroid lipofuscinosis due to progranulin gene mutation. Epilepsia. 2014;**55**(6):56-59

[27] Kleinberger G, Capell A, Haass C, Van Broeckhoven C. Mechanisms of granulin deficiency: Lessons from cellular and animal models. Molecular Neurobiology. 2013;**47**(1): 337-360

[28] Belcastro V, Siciliano V, Gregoretti F, Mithbaokar P, Dharmalingam G, Berlingieri S, et al. Transcriptional gene network inference from a massive dataset elucidates transcriptome organization and gene function. Nucleic Acids Research. 2011;**39**(20):8677-8688

[29] Capell A, Liebscher S, Fellerer K, Brouwers N, Willem M, Lammich S, et al. Rescue of progranulin deficiency associated with frontotemporal lobar degeneration by alkalizing reagents and inhibition of vacuolar ATPase. The Journal of Neuroscience. 2011;**31**(5):1885-1894

[30] Hu F, Padukkavidana T, Vægter CB, Brady OA, Zheng Y, Mackenzie IR, et al. Sortilin-mediated endocytosis determines levels of the frontotemporal dementia protein, progranulin. Neuron. 2010;**68**(4):654-667

[31] Zhou X, Sun L, de Oliveira FB, Qi X, Brown WJ, Smolka MB, et al. Prosaposin facilitates sortilin-independent lysosomal trafficking of progranulin. The Journal of Cell Biology. 2015;**210**(6):991-1002

[32] Zheng Y, Brady OA, Meng PS, Mao Y, Hu F. C-Terminus of Progranulin Interacts with the Beta-Propeller Region of Sortilin to Regulate Progranulin Trafficking. PLoS ONE. 2011;**6**(6):e21023.

[33] Pankiv S, Clausen TH, Lamark T, Brech A, Bruun JA, Outzen H, et al. p62/SQSTM1 binds directly to Atg8/LC3 to facilitate degradation of ubiquitinated protein aggregates by autophagy*[S]. The Journal of Biological Chemistry. 2007;**282**(33):24131-24145

[34] Tanaka Y, Chambers JK, Matsuwaki T, Yamanouchi K, Nishihara M. Possible involvement of lysosomal dysfunction in pathological changes of the brain in aged progranulin-deficient mice. Acta Neuropathologica Communications. 2014;**2**:78.

[35] Wils H, Kleinberger G, Pereson S, Janssens J, Capell A, Van Dam D, et al. Cellular ageing, increased mortality and FTLD-TDP-associated neuropathology in progranulin knockout mice. The Journal of Pathology. 2012;**228**(1):67-76

[36] Götzl JK, Mori K, Damme M, Fellerer K, Tahirovic S, Kleinberger G, et al. Common pathobiochemical hallmarks of progranulin-associated frontotemporal lobar degeneration and neuronal ceroid lipofuscinosis. Acta Neuropathologica. 2014;**127**(6):845-860

[37] Ahmed Z, Sheng H, Xu Y, Lin W-L, Innes AE, Gass J, et al. Accelerated lipofuscinosis and ubiquitination in granulin knockout mice suggest a role for progranulin in successful aging. American Journal of Pathology. 2010;**177**(1):311-324

[38] Yin F, Dumont M, Banerjee R, et al. Behavioral deficits and progressive neuropathology in progranulin-deficient mice: A mouse model of frontotemporal dementia. the Federation of American Societies for Experimental Biology journal. 2010;**24**(12):4639-4647

[39] Ghoshal N, Dearborn JT, Wozniak DF, Cairns NJ. Core features of frontotemporal dementia recapitulated in progranulin knockout mice. Neurobiology of Disease. 2012;**45** (1):395-408

[40] Petkau TL, Neal SJ, Milnerwood A, Mew A, Hill AM, Orban P, et al. Synaptic dysfunction in progranulin-deficient mice. Neurobiology of Disease. 2012;**45**(2):711-722

[41] Cruchaga C, Graff C, Chiang H-H, Wang J, Hinrichs AL, Spiegel N, et al. Association of TMEM106B gene polymorphism with age at onset in granulin mutation carriers and plasma granulin protein levels. Archives of Neurology. 2011;**68**(5):581-586

[42] Finch N, Carrasquillo MM, Baker M, Rutherford NJ, Coppola G, Dejesus-Hernandez M, et al. TMEM106B regulates progranulin levels and the penetrance of FTLD in GRN mutation carriers. Neurology. 2011;**76**(5):467-474

[43] Van Der Zee J, Van Langenhove T, Kleinberger G, Sleegers K, Engelborghs S, Vandenberghe R, et al. TMEM106B is associated with frontotemporal lobar degeneration in a clinically diagnosed patient cohort. Brain. 2011;**134**(3):808-815

[44] Van Deerlin VM, Sleiman PM, Martinez-Lage M, Chen-Plotkin A, Wang L-S, Graff-Radford NR, et al. Common variants at 7p21 are associated with frontotemporal lobar degeneration with TDP-43 inclusions. Nature Genetics. 2010;**42**(3):234-239

[45] Gallagher MD, Suh E, Grossman M, Elman L, McCluskey L, Van Swieten JC, et al. TMEM106B is a genetic modifier of frontotemporal lobar degeneration with C9orf72 hexanucleotide repeat expansions. Acta Neuropathologica. 2014;**127**(3):407-418

[46] Ash PEA, Bieniek KF, Gendron TF, Caulfield T, Lin WL, DeJesus-Hernandez M, et al. Unconventional translation of C9ORF72 GGGGCC expansion generates insoluble polypeptides specific to c9FTD/ALS. Neuron. 2013;**77**:639-646

[47] Chen-Plotkin AS, Unger TL, Gallagher MD, Bill E, Kwong LK, Volpicelli-Daley L, et al. TMEM106B, the risk gene for frontotemporal dementia, is regulated by the microRNA-132/212 cluster and affects progranulin pathways. The Journal of Neuroscience. 2012;**32** (33):11213-11227

[48] Brady OA, Zheng Y, Murphy K, Huang M, Hu F. The frontotemporal lobar degeneration risk factor, TMEM106B, regulates lysosomal morphology and function. Human Molecular Genetics. 2013;**22**(4):685-695

[49] Lang CM, Fellerer K, Schwenk BM, Kuhn PH, Kremmer E, Edbauer D, et al. Membrane orientation and subcellular localization of transmembrane protein 106B (TMEM106B), a major risk factor for frontotemporal lobar degeneration. The Journal of Biological Chemistry. 2012;**287**(23):19355-19365

[50] Stagi M, Klein ZA, Gould TJ, Bewersdorf J, Strittmatter SM. Lysosome size, motility and stress response regulated by fronto-temporal dementia modifier TMEM106B. Molecular and Cellular Neurosciences. 2014;**61**:226-240

[51] Schwenk BM, Lang CM, Hogl S, Tahirovic S, Orozco D, Rentzsch K, et al. The FTLD risk factor TMEM106B and MAP6 control dendritic trafficking of lysosomes. The European Molecular Biology Organization Journal. 2014;**33**(5):450-467

[52] Zhou X, Sun L, Brady OA, Murphy KA, Hu F. Elevated TMEM106B levels exaggerate lipofuscin accumulation and lysosomal dysfunction in aged mice with progranulin deficiency. Acta Neuropathologica Communications. 2017;**5**(1):9

[53] Skibinski G, Parkinson NJ, Brown JM, Chakrabarti L, Lloyd SL, Hummerich H, et al. Mutations in the endosomal ESCRTIII-complex subunit CHMP2B in frontotemporal dementia. Nature Genetics. 2005;**37**(8):806-808

[54] Urwin H, Josephs KA, Rohrer JD, MacKenzie IR, Neumann M, Authier A, et al. FUS pathology defines the majority of tau-and TDP-43-negative frontotemporal lobar degeneration. Acta Neuropathologica. 2010;**120**(1):33-41

[55] Parkinson N, Ince PG, Smith MO, Highley R, Skibinski G, Andersen PM, et al. ALS phenotypes with mutations in CHMP2B (charged multivesicular body protein 2B). Neurology. 2006;**67**(6):1074-1077

[56] Tsang HTH, Connell JW, Brown SE, Thompson A, Reid E, Sanderson CM. A systematic analysis of human CHMP protein interactions: Additional MIT domain-containing proteins bind to multiple components of the human ESCRT III complex. Genomics. 2006;**88**(3):333-346

[57] Fader CM, Colombo MI. Autophagy and multivesicular bodies: Two closely related partners. Cell Death and Differentiation. 2009;**16**(1):70-78

[58] Shim S, Kimpler LA, Hanson PI. Structure/function analysis of four core ESCRT-III proteins reveals common regulatory role for extreme C-terminal domain. Traffic. 2007;**8**(8): 1068-1079

[59] van der Zee J, Urwin H, Engelborghs S, Bruyland M, Vandenberghe R, Dermaut B, et al. CHMP2B C-truncating mutations in frontotemporal lobar degeneration are associated with an aberrant endosomal phenotype in vitro. Human Molecular Genetics. 2008;**17**(2):313-322

[60] Stuchell-Brereton MD, Skalicky JJ, Kieffer C, Karren MA, Ghaffarian S, Sundquist WI. ESCRT-III recognition by VPS4 ATPases. Nature. 2007;**449**(7163):740-744

[61] Obita T, Saksena S, Ghazi-Tabatabai S, Gill DJ, Perisic O, Emr SD, Williams RL. Structural basis for selective recognition of ESCRT-III by the AAA ATPase Vps4. Nature. 2007;**449**: 735-740

[62] Wollert T, Wunder C, Lippincott-schwartz J, Hurley JH. Membrane scission by the ESCRT-III complex. Nature. 2009;**458**(7235):172-177

[63] Lu Y, Zhang Z, Sun D, Sweeney ST, Gao FB. Syntaxin 13, a genetic modifier of mutant CHMP2B in frontotemporal dementia, is required for autophagosome maturation. Molecular Cell. 2013;**52**(2):264-271

[64] West RJH, Lu Y, Marie B, Gao FB, Sweeney ST. Rab8, POSH, and TAK1 regulate synaptic growth in a Drosophila model of frontotemporal dementia. The Journal of Cell Biology. 2015;**208**(7):931-947

[65] Filimonenko M, Stuffers S, Raiborg C, Yamamoto A, Malerød L, Fisher EMC, et al. Functional multivesicular bodies are required for autophagic clearance of protein aggregates associated with neurodegenerative disease. The Journal of Cell Biology. 2007; **179**(3):485-500

[66] Lee JA, Liu L, Gao FB. Autophagy defects contribute to neurodegeneration induced by dysfunctional ESCRT-III. Autophagy. 2009;**5**:1070-1072

[67] Ghazi-Noori S, Froud KE, Mizielinska S, Powell C, Smidak M, Fernandez De Marco M, et al. Progressive neuronal inclusion formation and axonal degeneration in CHMP2B mutant transgenic mice. Brain. 2012;**135**(3):819-832

[68] Krasniak CS, Ahmad ST. The role of CHMP2BIntron5 in autophagy and frontotemporal dementia. Brain Research. 2016;**1649**:151-157

[69] Vernay A, Therreau L, Blot B, Risson V, Dirrig-Grosch S, Waegaert R, et al. A transgenic mouse expressing CHMP2Bintron5 mutant in neurons develops histological and behavioural features of amyotrophic lateral sclerosis and frontotemporal dementia. Human Molecular Genetics. 2016;**25**(15):3341-3360

[70] Nielsen TT, Mizielinska S, Hasholt L, Isaacs AM, Nielsen JE. Reversal of pathology in CHMP2B-mediated frontotemporal dementia patient cells using RNA interference. The Journal of Gene Medicine. 2012;**14**(8):521-529

[71] Clayton EL, Mizielinska S, Edgar JR, Nielsen TT, Marshall S, Norona FE, et al. Frontotemporal dementia caused by CHMP2B mutation is characterised by neuronal lysosomal storage pathology. Acta Neuropathologica. 2015;**130**(4):511-523

[72] Watts GDJ, Wymer J, Kovach MJ, Mehta SG, Mumm S, Darvish D, et al. Inclusion body myopathy associated with Paget disease of bone and frontotemporal dementia is caused by mutant valosin-containing protein. Nature Genetics. 2004;**36**(4):377-381

[73] Guyant-Maréchal L, Laquerrière A, Duyckaerts C, Dumanchin C, Bou J, Dugny F, et al. Valosin-containing protein gene mutations: Clinical and neuropathologic features. Neurology. 2006;**67**(4):644-651

[74] Forman MS, Mackenzie IR, Cairns NJ, Swanson E, Boyer PJ, Drachman DA, et al. Novel ubiquitin neuropathology in frontotemporal dementia with valosin-containing protein gene mutations. Journal of Neuropathology and Experimental Neurology. 2006;**65**(6):571-581

[75] Schröder R, Watts GDJ, Mehta SG, Evert BO, Broich P, Fließbach K, et al. Mutant valosin-containing protein causes a novel type of frontotemporal dementia. Annals of Neurology. 2005;**57**(3):457-461

[76] Neumann M, Mackenzie IR, Cairns NJ, Boyer PJ, Markesbery WR, Smith CD, et al. TDP-43 in the ubiquitin pathology of frontotemporal dementia with VCP gene mutations. Journal of Neuropathology and Experimental Neurology. 2007;**66**(2):152-157

[77] Johnson JO, Mandrioli J, Benatar M, Abramzon Y, Van Deerlin VM, Trojanowski JQ, et al. Exome sequencing reveals VCP mutations as a cause of familial ALS. Neuron. 2010;**68**(5):857-864

[78] Gonzalez MA, Feely SM, Speziani F, Strickland AV, Danzi M, Bacon C, et al. A novel mutation in VCP causes Charcot-Marie-Tooth Type 2 disease. Brain. 2014;**137**(11):2897-2902

[79] Jentsch S, Rumpf S. Cdc48 (p97): A "molecular gearbox" in the ubiquitin pathway? Trends in Biochemical Sciences. 2007;**32**(1):6-11

[80] Rabinovich E, Kerem A, Fröhlich K-U, Diamant N, Bar-Nun S. AAA-ATPase p97/Cdc48p, a cytosolic chaperone required for endoplasmic reticulum-associated protein degradation. Molecular and Cellular Biology. 2002;**22**(2):626-634

[81] Song C, Wang Q, Song C, Rogers TJ. Valosin-containing protein (VCP/p97) is capable of unfolding polyubiquitinated proteins through its ATPase domains. Biochemical and Biophysical Research Communications. 2015;**463**(3):453-457

[82] Lim PJ, Danner R, Liang J, Doong H, Harman C, Srinivasan D, et al. Ubiquilin and p97/VCP bind erasin, forming a complex involved in ERAD. The Journal of Cell Biology. 2009;**187**(2):201-217

[83] Ye Y, Meyer HH, Rapoport TA. The AAA ATPase Cdc48/p97 and its partners transport proteins from the ER into the cytosol. Nature. 2001;**414**(6864):652-656

[84] Pleasure IT, Black MM, Keen JH. Valosin-containing protein, VCP, is a ubiquitous clathrin-binding protein. Nature. 1993;**365**(6445):459-462

[85] Ramanathan HN, Ye Y. The p97 ATPase associates with EEA1 to regulate the size of early endosomes. Cell Research. 2012;**22**(2):346-359

[86] Ju JS, Fuentealba RA, Miller SE, Jackson E, Piwnica-Worms D, Baloh RH, et al. Valosin-containing protein (VCP) is required for autophagy and is disrupted in VCP disease. The Journal of Cell Biology. 2009;**187**(6):875-888

[87] Tresse E, Salomons FA, Vesa J, Bott LC, Kimonis V, Yao TP, et al. VCP/p97 is essential for maturation of ubiquitin-containing autophagosomes and this function is impaired by mutations that cause IBMPFD. Autophagy. 2010;**6**(2):217-227

[88] Ritz D, Vuk M, Kirchner P, Bug M, Schütz S, Hayer A, et al. Endolysosomal sorting of ubiquitylated caveolin-1 is regulated by VCP and UBXD1 and impaired by VCP disease mutations. Nature Cell Biology. 2011;**13**(9):1116-1123

[89] Badadani M, Nalbandian A, Watts GD, et al. VCP Associated Inclusion Body Myopathy and Paget Disease of Bone Knock-In Mouse Model Exhibits Tissue Pathology Typical of Human Disease. PLoS ONE. 2010;**5**(10):e13183.

[90] Custer SK, Neumann M, Lu H, Wright AC, Taylor JP. Transgenic mice expressing mutant forms VCP/p97 recapitulate the full spectrum of IBMPFD including degeneration in muscle, brain and bone. Human Molecular Genetics. 2010;**19**(9):1741-1755

[91] Müller JMM, Deinhardt K, Rosewell I, Warren G, Shima DT. Targeted deletion of p97 (VCP/CDC48) in mouse results in early embryonic lethality. Biochemical and Biophysical Research Communications. 2007;**354**(2):459-465

[92] Buchan JR, Kolaitis RM, Taylor JP, Parker R. XEukaryotic stress granules are cleared by autophagy and Cdc48/VCP function. Cell. 2013;**153**(7):1461-1474

[93] Weihl CC, Temiz P, Miller SE, Watts G, Smith C, Forman M, et al. TDP-43 accumulation in inclusion body myopathy muscle suggests a common pathogenic mechanism with frontotemporal dementia. Journal of Neurology Neurosurgery and Psychiatry. 2008;**79**(10):1186-1189

[94] Fecto F. Mutations in familial and sporadic amyotrophic lateral sclerosis. Archives of Neurology. 2011;**68**(11):1440-1446

[95] Deng H-X, Chen W, Hong S-T, Boycott KM, Gorrie GH, Siddique N, et al. Mutations in UBQLN2 cause dominant X-linked juvenile and adult-onset ALS and ALS/dementia. Nature. 2011;**477**(7363):211-215

[96] Freischmidt A, Müller K, Ludolph AC, Weishaupt JH, Andersen PM. Association of mutations in TBK1 with sporadic and familial amyotrophic lateral sclerosis and frontotemporal dementia. JAMA Neurology. 2017;**74**(1):110-113

[97] Weishaupt JH, Waibel S, Birve A, Volk AE, Mayer B, Meyer T, et al. A novel optineurin truncating mutation and three glaucoma-associated missense variants in patients with familial amyotrophic lateral sclerosis in Germany. Neurobiology of Aging. 2013;**34**(5):1516.e9-1516.e15

[98] Williams KL, McCann EP, Fifita JA, Zhang K, Duncan EL, Leo PJ, et al. Novel TBK1 truncating mutation in a familial amyotrophic lateral sclerosis patient of Chinese origin. Neurobiology of Aging. 2015;**36**(12):3334.e1-3334.e5

[99] Maruyama H, Morino H, Ito H, Izumi Y, Kato H, Watanabe Y, et al. Mutations of optineurin in amyotrophic lateral sclerosis. Nature. 2010;**465**(7295):223-236

[100] Kovacs GG, van der Zee J, Hort J, Kristoferitsch W, Leitha T, Höftberger R, et al. Clinicopathological description of two cases with SQSTM1 gene mutation associated with frontotemporal dementia. Neuropathology. 2016;**36**(1):27-38

[101] Pottier C, Bieniek KF, Finch NC, van de Vorst M, Baker M, Perkersen R, et al. Whole-genome sequencing reveals important role for TBK1 and OPTN mutations in frontotemporal lobar degeneration without motor neuron disease. Acta Neuropathologica. 2015;**130**(1):77-92

[102] Arai T, Nonaka T, Hasegawa M, Akiyama H, Yoshida M, Hashizume Y, et al. Neuronal and glial inclusions in frontotemporal dementia with or without motor neuron disease are immunopositive for p62. Neuroscience Letters. 2003;**342**(1-2):41-44

[103] Kuusisto E, Kauppinen T, Alafuzoff I. Use of p62/SQSTM1 antibodies for neuropathological diagnosis. Neuropathology and Applied Neurobiology. 2008;**34**(2):169-180

[104] Nakano T, Nakaso K, Nakashima K, Ohama E. Expression of ubiquitin-binding protein p62 in ubiquitin-immunoreactive intraneuronal inclusions in amyotrophic lateral sclerosis with dementia: Analysis of five autopsy cases with broad clinicopathological spectrum. Acta Neuropathologica. 2004;**107**(4):359-364

[105] Korolchuk VI, Menzies FM, Rubinsztein DC. A novel link between autophagy and the ubiquitin-proteasome system. Autophagy. 2009;**5**(6):862-863

[106] Seibenhener M, Babu J. Sequestosome 1/p62 Is a polyubiquitin chain binding protein involved in ubiquitin proteasome degradation. Molecular and Cellular Biology. 2004;**24**(18):8055-8068

[107] Rea SL, Walsh JP, Layfield R, Ratajczak T, Xu Jiake J. New insights into the role of sequestosome 1/p62 mutant proteins in the pathogenesis of paget's disease of bone. Endocrine Reviews. 2013;**34**(4):501-524

[108] Rogov V, Dötsch V, Johansen T, Kirkin V. Interactions between autophagy receptors and Ubiquitin-like proteins form the molecular basis for selective autophagy. Molecular Cell. 2014;**53**(2):167-178

[109] Johansen T, Lamark T. Selective autophagy mediated by autophagic adapter proteins. Autophagy. 2011;**7**(3):279-296

[110] Matsumoto G, Wada K, Okuno M, Kurosawa M, Nukina N. Serine 403 phosphorylation of p62/SQSTM1 regulates selective autophagic clearance of ubiquitinated proteins. Molecular Cell. 2011;**44**(2):279-289

[111] Pilli M, Arko-Mensah J, Ponpuak M, Roberts E, Master S, Mandell MA, et al. TBK-1 Promotes Autophagy-Mediated antimicrobial defense by controlling autophagosome maturation. Immunity. 2012;**37**(2):223-234

[112] Lim J, Lachenmayer ML, Wu S, Liu W, Kundu M, Wang R, et al. Proteotoxic stress induces phosphorylation of p62/SQSTM1 by ULK1 to regulate selective autophagic clearance of protein aggregates. PLoS Genetics. 2015;**11**(2):e1004987

[113] Ro SH, Semple IA, Park H, Park H, Park HW, Kim M, et al. Sestrin2 promotes Unc-51-like kinase 1 mediated phosphorylation of p62/sequestosome-1. The FEBS Journal. 2014;**281**(17):3816-3827

[114] Paine MG, Babu JR, Seibenhener ML, Wooten MW. Evidence for p62 aggregate formation: Role in cell survival. The Federation of European Biochemical Societies Letters. 2005;**579**(22):5029-5034

[115] Bjørkøy G, Lamark T, Brech A, Outzen H, Perander M, Øvervatn A, et al. p62/SQSTM1 forms protein aggregates degraded by autophagy and has a protective effect on huntingtin-induced cell death. The Journal of Cell Biology. 2005;**171**(4):603-614

[116] Komatsu M, Waguri S, Koike M, Sou Y shin, Ueno T, Hara T, et al. Homeostatic levels of p62 control cytoplasmic inclusion body formation in autophagy-deficient mice. Cell. 2007;**131**(6):1149-1163

[117] Kwok CT, Morris A, de Belleroche JS. Sequestosome-1 (SQSTM1) sequence variants in ALS cases in the UK: Prevalence and coexistence of SQSTM1 mutations in ALS kindred with PDB. European Journal of Human Genetics. 2014;**22**(4):492-496

[118] Teyssou E, Takeda T, Lebon V, Boillée S, Doukouré B, Bataillon G, et al. Mutations in SQSTM1 encoding p62 in amyotrophic lateral sclerosis: Genetics and neuropathology. Acta Neuropathologica. 2013;**125**(4):511-522

[119] Rubino E, Chio A, Rogaeva E, Galimberti D, Bruni AC, St PH. SQSTM1 mutations in frontotemporal lobar degeneration and amyotrophic lateral sclerosis. Neurology. 2012;**79**(15):1556-62

[120] Haack TB, Ignatius E, Calvo-Garrido J, et al. Absence of the Autophagy Adaptor SQSTM1/p62 Causes Childhood-Onset Neurodegeneration with Ataxia, Dystonia, and Gaze Palsy. American Journal of Human Genetics. 2016;**99**(3):735-743.

[121] Babu JR, Geetha T, Wooten MW. Sequestosome 1/p62 shuttles polyubiquitinated tau for proteasomal degradation. Journal of Neurochemistry. 2005;**94**(1):192-203

[122] Zhang Y-J, Gendron TF, Xu Y-F, Ko L-W, Yen S-H, Petrucelli L. Phosphorylation regulates proteasomal-mediated degradation and solubility of TAR DNA binding protein-43 C-terminal fragments. Molecular Neurodegeneration. 2010;**5**:33

[123] Ko HS, Uehara T, Tsuruma K, Nomura Y. Ubiquilin interacts with ubiquitylated proteins and proteasome through its ubiquitin-associated and ubiquitin-like domains. The Federation of European Biochemical Societies Letters. 2004;**566**(1-3):110-114

[124] N'Diaye E-N, Kajihara KK, Hsieh I, Morisaki H, Debnath J, Brown EJ. PLIC proteins or ubiquilins regulate autophagy-dependent cell survival during nutrient starvation. The European Molecular Biology Organization Reports. 2009;**10**(2):173-179

[125] Rothenberg C, Srinivasan D, Mah L, Kaushik S, Peterhoff CM, Ugolino J, et al. Ubiquilin functions in autophagy and is degraded by chaperone-mediated autophagy. Human Molecular Genetics. 2010;**19**(16):3219-3232

[126] Cassel JA, Reitz AB. Ubiquilin-2 (UBQLN2) binds with high affinity to the C-terminal region of TDP-43 and modulates TDP-43 levels in H4 cells: Characterization of inhibition by nucleic acids and 4-aminoquinolines. Biochimica et Biophysica Acta - Proteins Proteomics. 2013;**1834**(6):964-971

[127] Majcher V, Goode A, James V, Layfield R. Autophagy receptor defects and ALS-FTLD. Molecular and Cellular Neurosciences. 2015;**66**(Part A):43-52

[128] Gilpin KM, Chang L, Monteiro MJ. ALS-linked mutations in ubiquilin-2 or hnRNPA1 reduce interaction between ubiquilin-2 and hnRNPA1. Human Molecular Genetics. 2015;**24**(9):2565-2577

[129] Wu Q, Liu M, Huang C, Liu X, Huang B, Li N, et al. Pathogenic Ubqln2 gains toxic properties to induce neuron death. Acta Neuropathologica. 2014;**129**(3):417-428

[130] Le NTT, Chang L, Kovlyagina I, Georgiou P, Safren N, Braunstein KE, et al. Motor neuron disease, TDP-43 pathology, and memory deficits in mice expressing ALS-FTD-linked UBQLN2 mutations. In: Proceedings of the National Academy of Sciences of the United States of America. 2016;**113**(47):E7580-E7589.

[131] Ceballos-Diaz C, Rosario AM, Park H-J, Chakrabarty P, Sacino A, Cruz PE, et al. Viral expression of ALS-linked ubiquilin-2 mutants causes inclusion pathology and behavioral deficits in mice. Molecular Neurodegeneration. 2015;**10**(1):25

[132] Huang, B., Wu, Q., Zhou, H., Huang, C. and Xia, X.-G. Increased Ubqln2 expression causes neuron death in transgenic rats. J. Neurochem., 2016;**139**: 285-293.

[133] Walters KJ, Kleijnen MF, Goh AM, Wagner G, Howley PM. Structural studies of the interaction between ubiquitin family proteins and proteasome subunit S5a. Biochemistry. 2002;**41**(6):1767-1777

[134] Hjerpe R, Bett JS, Keuss MJ, Solovyova A, McWilliams TG, Johnson C, et al. UBQLN2 mediates Autophagy-Independent protein aggregate clearance by the proteasome. Cell. 2016;**166**(4):935-949

[135] Osaka M, Ito D, Suzuki N. Disturbance of proteasomal and autophagic protein degradation pathways by amyotrophic lateral sclerosis-linked mutations in ubiquilin 2. Biochemical and Biophysical Research Communications. 2016;**472**(2):324-331

[136] Xia Y, Yan LH, Huang B, Liu M, Liu X, Huang C. Pathogenic mutation of UBQLN2 impairs its interaction with UBXD8 and disrupts endoplasmic reticulum-associated protein degradation. Journal of Neurochemistry. 2014;**129**(1):99-106

[137] Iguchi Y, Katsuno M, Ikenaka K, Ishigaki S, Sobue G. Amyotrophic lateral sclerosis: An update on recent genetic insights. Journal of Neurology. 2013;**260**(11):2917-2927

[138] Wild P, Farhan H, McEwan DG, Wagner S, Rogov V V., Brady NR, et al. Phosphorylation of the autophagy. Science (80-). 2011;**333**(July):228-233

[139] Wong YC, Holzbaur ELF. Optineurin is an autophagy receptor for damaged mitochondria in parkin-mediated mitophagy that is disrupted by an ALS-linked mutation. Proceedings of the National Academy of Sciences of the United States of America. 2014;**111**(42):E4439-4448

[140] Korac J, Schaeffer V, Kovacevic I, Clement AM, Jungblut B, Behl C, et al. Ubiquitin-independent function of optineurin in autophagic clearance of protein aggregates. Journal of Cell Science. 2013;**126**(2):580-592

[141] Liu Z, Chen P, Gao H, Gu Y, Yang J, Peng H, et al. Ubiquitylation of autophagy receptor optineurin by HACE1 activates selective autophagy for tumor suppression. Cancer Cell. 2014;**26**(1):106-120

[142] Morton S, Hesson L, Peggie M, Cohen P. Enhanced binding of TBK1 by an optineurin mutant that causes a familial form of primary open angle glaucoma. The Federation of European Biochemical Societies Letters. 2008;**582**(6):997-1002

[143] Chalasani MLS, Kumari A, Radha V, Swarup G. E50K-OPTN-Induced Retinal Cell Death Involves the Rab GTPase-Activating Protein, TBC1D17 Mediated Block in Autophagy. PLoS ONE. 2014;**9**(4):e95758.

[144] Borghero G, Pugliatti M, Marrosu F, Marrosu MG, Murru MR, Floris G, et al. TBK1 is associated with ALS and ALS-FTD in Sardinian patients. Neurobiology of Aging. 2016;**43**:1-5

[145] Gijselinck I, Van Mossevelde S, van der Zee J, Sieben A, Philtjens S, Heeman B, et al. Loss of *TBK1* is a frequent cause of frontotemporal dementia in a Belgian cohort. Neurology. 2015;**85**(24):2116-2125

[146] Le Ber I, De Septenville A, Millecamps S, Camuzat A, Caroppo P, Couratier P, et al. TBK1 mutation frequencies in French frontotemporal dementia and amyotrophic lateral sclerosis cohorts. Neurobiology of Aging. 2015;**36**(11):3116.e5-3116.e8

[147] Tsai PC, Liu YC, Lin KP, Liu YT, Liao YC, Hsiao CT, et al. Mutational analysis of TBK1 in Taiwanese patients with amyotrophic lateral sclerosis. Neurobiology of Aging. 2016;**40**:191.e11-6

[148] Oakes JA, Davies MC, Collins MO. TBK1: a new player in ALS linking autophagy and neuroinflammation. Molecular Brain. 2017;**10**:5.

[149] Minegishi Y, Nakayama M, Iejima D, Kawase K, Iwata T. Significance of optineurin mutations in glaucoma and other diseases. Progress in Retinal and Eye Research. 2016;**55**:149-181

[150] Moore AS, Holzbaur ELF. Dynamic recruitment and activation of ALS-associated TBK1 with its target optineurin are required for efficient mitophagy. Proceedings of the National Academy of Sciences of the United States of America. 2016;**113**(24):E3349–E3358.

[151] Dejesus-hernandez M, Mackenzie IR, Boeve BF, Boxer AL, Baker M, Rutherford NJ, et al. Supplemental information expanded GGGGCC hexanucleotide repeat in noncoding region of C9ORF72 causes chromosome 9p-Linked FTD and ALS. Neuron. 2011;**72**:245-256

[152] Renton AE, Majounie E, Waite A, Simón-Sánchez J, Rollinson S, Gibbs JR, et al. A hexanucleotide repeat expansion in C9ORF72 is the cause of chromosome 9p21-linked ALS-FTD. Neuron. 2011;**72**:257-268

[153] Gijselinck I, Van Langenhove T, van der Zee J, Sleegers K, Philtjens S, Kleinberger G, et al. A C9orf72 promoter repeat expansion in a Flanders-Belgian cohort with disorders of the frontotemporal lobar degeneration-amyotrophic lateral sclerosis spectrum: A gene identification study. The Lancet Neurology. 2012;**11**:54-65

[154] Majounie E, Renton AE, Mok K, Dopper EGP, Waite A, Rollinson S, et al. Frequency of the C9orf72 hexanucleotide repeat expansion in patients with amyotrophic lateral sclerosis and frontotemporal dementia: A cross-sectional study. The Lancet Neurology. 2012;**11**(4):323-330

[155] van der Zee J, Gijselinck I, Dillen L, Van Langenhove T, Theuns J, Engelborghs S, et al. A Pan-European study of the C9orf72 repeat associated with FTLD: Geographic prevalence, genomic instability, and intermediate repeats. Human Mutation. 2013;**34**(2): 363-373

[156] Al-Sarraj S, King A, Troakes C, Smith B, Maekawa S, Bodi I, et al. P62 positive, TDP-43 negative, neuronal cytoplasmic and intranuclear inclusions in the cerebellum and hippocampus define the pathology of C9orf72-linked FTLD and MND/ALS. Acta Neuropathologica. 2011;**122**(6):691-702

[157] Brettschneider J, Van Deerlin VM, Robinson JL, Kwong L, Lee EB, Ali YO, et al. Pattern of ubiquilin pathology in ALS and FTLD indicates presence of C9ORF72 hexanucleotide expansion. Acta Neuropathologica. 2012;**123**(6):825-839

[158] King A, Maekawa S, Bodi I, Troakes C, Al-Sarraj S. Ubiquitinated, p62 immunopositive cerebellar cortical neuronal inclusions are evident across the spectrum of TDP-43 proteinopathies but are only rarely additionally immunopositive for phosphorylation-dependent TDP-43. Neuropathology. 2011;**31**(3):239-249

[159] Mahoney CJ, Beck J, Rohrer JD, Lashley T, Mok K, Shakespeare T, et al. Frontotemporal dementia with the C9ORF72 hexanucleotide repeat expansion: Clinical, neuroanatomical and neuropathological features. Brain. 2012;**135**(3):736-750

[160] Mori K, Weng S-M, Arzberger T, May S, Rentzsch K, Kremmer E, et al. The C9orf72 GGGGCC repeat is translated into aggregating dipeptide-repeat proteins in FTLD/ALS. Science. 2013;**339**:1335-1338

[161] Lagier-Tourenne C, Baughn M, Rigo F, Sun S, Liu P, Li H-R, et al. Targeted degradation of sense and antisense C9orf72 RNA foci as therapy for ALS and frontotemporal degeneration. Proceedings of the National Academy of Sciences of the United States of America. 2013;**110**(47):E45304539

[162] Mizielinska S, Lashley T, Norona FE, Clayton EL, Ridler CE, Fratta P, et al. C9orf72 frontotemporal lobar degeneration is characterised by frequent neuronal sense and antisense RNA foci. Acta Neuropathologica. 2013;**126**(6):845-857

[163] Gendron TF, Bieniek KF, Zhang YJ, Jansen-West K, Ash PEA, Caulfield T, et al. Antisense transcripts of the expanded C9ORF72 hexanucleotide repeat form nuclear RNA foci and undergo repeat-associated non-ATG translation in c9FTD/ALS. Acta Neuropathologica. 2013;**126**(6):829-844

[164] Freibaum BD, Lu Y, Lopez-Gonzalez R, Kim NC, Almeida S, Lee K-H, et al. GGGGCC repeat expansion in C9orf72 compromises nucleocytoplasmic transport. Nature. 2015;**525**(7567):129-133

[165] Gendron TF, Belzil V V, Zhang Y-J, Petrucelli L. Mechanisms of toxicity in C9FTLD/ALS. Acta Neuropathologica. 2014;**127**(3):359-376

[166] Zhang K, Donnelly CJ, Haeusler AR, Grima JC, Machamer JB, Steinwald P, et al. The C9orf72 repeat expansion disrupts nucleocytoplasmic transport. Nature. 2015; **525**(7567):56-61

[167] Khosravi B, Hartmann H, May S, Möhl C, Ederle H, Michaelsen M, et al. Cytoplasmic poly-GA aggregates impair nuclear import of TDP-43 in *C9orf72* ALS/FTLD. Human Molecular Genetics. 2017;**26**(4):790-800

[168] Zhang Y-J, Gendron TF, Grima JC, Sasaguri H, Jansen-West K, Xu Y-F, et al. C9ORF72 poly(GA) aggregates sequester and impair HR23 and nucleocytoplasmic transport proteins. Nature Neuroscience. 2016;**19**(5):668-677

[169] Lee KH, Zhang P, Kim HJ, Mitrea DM, Sarkar M, Freibaum BD, et al. C9orf72 dipeptide repeats impair the assembly, dynamics, and function of Membrane-Less organelles. Cell. 2016;**167**(3):774-788.e17

[170] Fratta P, Poulter M, Lashley T, Rohrer JD, Polke JM, Beck J, et al. Homozygosity for the C9orf72 GGGGCC repeat expansion in frontotemporal dementia. Acta Neuropathologica. 2013;**126**(3):401-409

[171] van Blitterswijk M, Gendron TF, Baker MC, DeJesus-Hernandez M, Finch NCA, Brown PH, et al. Novel clinical associations with specific C9ORF72 transcripts in patients with repeat expansions in C9ORF72. Acta Neuropathologica. 2015;**130**(6):863-876

[172] Waite AJ, Bäumer D, East S, et al. Reduced C9orf72 protein levels in frontal cortex of amyotrophic lateral sclerosis and frontotemporal degeneration brain with the C9ORF72 hexanucleotide repeat expansion. Neurobiology of Aging. 2014;**35**(7):1779.e5-1779.e13.

[173] Therrien M, Rouleau GA, Dion PA, Parker JA. Deletion of C9ORF72 Results in Motor Neuron Degeneration and Stress Sensitivity in C. elegans. PLoS ONE. 2013;**8**(12):e83450.

[174] Ciura S, Lattante S, Le Ber I, Latouche M, Tostivint H, Brice A, et al. Loss of function of C9orf72 causes motor deficits in a zebrafish model of amyotrophic lateral sclerosis. Annals of Neurology. 2013;**74**:180-187

[175] Koppers M, Blokhuis AM, Westeneng HJ, Terpstra ML, Zundel CAC, Vieira De Sá R, et al. C9orf72 ablation in mice does not cause motor neuron degeneration or motor deficits. Annals of Neurology. 2015;**78**(3):426-438

[176] Burberry A, Suzuki N, Wang JY, Moccia R, Mordes DA, Stewart MH, et al. Loss-of-function mutations in the C9ORF72 mouse ortholog cause fatal autoimmune disease. Science Translational Medicine. 2016;**8**(347):347ra93

[177] ORourke JG, Bogdanik L, Yanez A, Lall D, Wolf AJ, Muhammad AKMG, et al. C9orf72 is required for proper macrophage and microglial function in mice. Science (80-). 2016;**351**(6279):1324-1329

[178] Atanasio A, Decman V, White D, Ramos M, Ikiz B, Lee H-C, et al. C9orf72 ablation causes immune dysregulation characterized by leukocyte expansion, autoantibody production, and glomerulonephropathy in mice. Scientific Reports. 2016;**6**(November 2015):23204

[179] Sullivan PM, Zhou X, Robins AM, Paushter DH, Kim D, Smolka MB, et al. The ALS/FTLD associated protein C9orf72 associates with SMCR8 and WDR41 to regulate the autophagy-lysosome pathway. Acta Neuropathologica Communications. 2016;**4**(1):51

[180] Sudria-Lopez E, Koppers M, de Wit M, van der Meer C, Westeneng HJ, Zundel CAC, et al. Full ablation of C9orf72 in mice causes immune system-related pathology and neoplastic events but no motor neuron defects. Acta Neuropathologica. 2016;**132**(1):145-147

[181] Jiang J, Zhu Q, Gendron TF, Saberi S, McAlonis-Downes M, Seelman A, et al. Gain of toxicity from ALS/FTD-Linked repeat expansions in C9ORF72 is alleviated by antisense oligonucleotides targeting GGGGCC-Containing RNAs. Neuron. 2016;**90**(3):535-550

[182] Farg MA, Sundaramoorthy V, Sultana JM, Yang S, Atkinson RAK, Levina V, et al. C9ORF72, implicated in amytrophic lateral sclerosis and frontotemporal dementia, regulates endosomal trafficking. Human Molecular Genetics. 2014;**23**(13):3579-3595

[183] Sellier C, Campanari M-L, Julie Corbier C, Gaucherot A, Kolb-Cheynel I, Oulad-Abdelghani M, et al. Loss of C9ORF72 impairs autophagy and synergizes with polyQ Ataxin-2 to induce motor neuron dysfunction and cell death. The European Molecular Biology Organization Journal. 2016;**35**(Icm):1-22

[184] Yang M, Liang C, Swaminathan K, Herrlinger S, Lai F, Shiekhattar R, et al. A C9ORF72/SMCR8-containing complex regulates ULK1 and plays a dual role in autophagy. Science Advances. 2016;**2**(9):e1601167-e1601167

[185] Webster CP, Smith EF, Bauer CS, Moller A, Guillaume M, Ferraiuolo L, et al. The C9orf72 protein interacts with Rab 1 a and the ULK 1 complex to regulate initiation of autophagy. The European Molecular Biology Organization Journal. 2016;**35**(15):1-21

[186] Ugolino J, Ji YJ, Conchina K, Chu J, Nirujogi RS, Pandey A, et al. Loss of C9orf72 Enhances Autophagic Activity via Deregulated mTOR and TFEB Signaling. PLoS Genetics. 2016;**12**(11):e1006443. doi:10.1371/journal.pgen.1006443.

[187] Amick J, Roczniak-Ferguson A, Ferguson SM. C9orf72 binds SMCR8, localizes to lysosomes, and regulates mTORC1 signaling. Molecular Biology of the Cell. 2016;**27**(20):3040-3051.

[188] Maharjan N, Kunzli C, Buthey K SS. C9ORF72 Regulates stress granule formation and its deficiency impairs stress granule assembly, hypersensitizing cells to stress. Human Molecular Genetics. 2016;**25**(15):3341-3360

[189] Sivadasan R, Hornburg D, Drepper C, Frank N, Jablonka S, Hansel A, et al. C9ORF72 interaction with cofilin modulates actin dynamics in motor neurons. Nature Neuroscience. 2016;**19**(12):1610-1618

[190] Snowden JS, Hu Q, Rollinson S, Halliwell N, Robinson A, Davidson YS, et al. The most common type of FTLD-FUS (aFTLD-U) is associated with a distinct clinical form of frontotemporal dementia but is not related to mutations in the FUS gene. Acta Neuropathologica. 2011;**122**(1):99-110

[191] Lagier-Tourenne C, Polymenidou M, Cleveland DW. TDP-43 and FUS/TLS: Emerging roles in RNA processing and neurodegeneration. Human Molecular Genetics. 2010;**19** (R1):46-64

[192] Bosco DA, Lemay N, Ko HK, Zhou H, Burke C, Kwiatkowski TJ, et al. Mutant FUS proteins that cause amyotrophic lateral sclerosis incorporate into stress granules. Human Molecular Genetics. 2010;**19**(21):4160-4175

[193] Thomas M, Alegre-Abarrategui J, Wade-Martins R. RNA dysfunction and aggrephagy at the centre of an amyotrophic lateral sclerosis/frontotemporal dementia disease continuum. Brain. 2013;**136**(5):1345-1360

[194] Cruts M, Theuns J, Van Broeckhoven C. Locus-specific mutation databases for neurodegenerative brain diseases. Human Mutation. 2012;**33**(9):1340-1344

[195] Kabashi E, Valdmanis P, Dion P, Spiegelman D, McConkey BJ, Vande Velde C, Bouchard JP, Lacomblez L, Pochigaeva K, Salachas F, Pradat PF, Camu W, Meininger V, Dupre N, Rouleau GA. TARDBP mutations in individuals with sporadic and familial amyotrophoc lateral sclerosis. Nature Genetics. 2008;**40**(5):572-574

[196] Huang C-C, Bose JK, Majumder P, Lee K-H, Huang J-TJ, Huang JK, et al. Metabolism and mis-metabolism of the neuropathological signature protein TDP-43. Journal of Cell Science. 2014;**127**(Pt 14):3024-3038

[197] Xia Q, Wang H, Hao Z, Fu C, Hu Q, Gao F, et al. TDP-43 loss of function increases TFEB activity and blocks autophagosome–lysosome fusion. The European Molecular Biology Organization Journal. 2016;**35**(2):121-142

[198] Monahan Z, Shewmaker F, Pandey UB. Stress granules at the intersection of autophagy and ALS. Brain Research. 2016;**1649**:189-200

[199] Bose JK, Huang CC, Shen CKJ. Regulation of autophagy by neuropathological protein TDP-43. The Journal of Biological Chemistry. 2011;**286**(52):44441-44448

[200] Molliex A, Temirov J, Lee J, Coughlin M, Kanagaraj AP, Kim HJ, et al. Phase separation by low complexity domains promotes stress granule assembly and drives pathological fibrillization. Cell. 2015;**163**(1):123-133

[201] Murakami T, Qamar S, Lin JQ, Schierle GSK, Rees E, Miyashita A, et al. ALS/FTD Mutation-Induced phase transition of FUS liquid droplets and reversible hydrogels into irreversible hydrogels impairs RNP granule function. Neuron. 2015;**88**(4):678-690

[202] Li YR, King OD, Shorter J, Gitler AD. Stress granules as crucibles of ALS pathogenesis. The Journal of Cell Biology. 2013;**201**(3):361-372

[203] Taylor JP, Brown RH, Cleveland DW. Decoding ALS: From genes to mechanism. Nature. 2016;**539**(7628):197-206

[204] Zhou Y, Liu S, Liu G, Öztürk A, Hicks GG. ALS-Associated FUS Mutations Result in Compromised FUS Alternative Splicing and Autoregulation. PLoS Genetics. 2013;**9**(10):e1003895.

[205] Ryu HH, Jun MH, Min KJ, Jang DJ, Lee YS, Kim HK, et al. Autophagy regulates amyotrophic lateral sclerosis-linked fused in sarcoma-positive stress granules in neurons. Neurobiology of Aging. 2014;**35**(12):2822-2831

[206] Gal J, Zhang J, Kwinter DM, Zhai J, Jia H, Jia J, et al. Nuclear localization sequence of FUS and induction of stress granules by ALS mutants. Neurobiology of Aging. 2011;**32**(12):2323.e27-2323.e40

[207] Ling S-C, Polymenidou M, Cleveland DW. Converging mechanisms in ALS and FTD: Disrupted RNA and protein homeostasis. Neuron. 2013;**79**(3):416-438

[208] Dormann D, Haass C. TDP-43 and FUS: A nuclear affair. Trends in Neurosciences. 2011; **34**(7):339-348

[209] Patel A, Lee HO, Jawerth L, Maharana S, Jahnel M, Hein MY, et al. A Liquid-to-Solid phase transition of the ALS protein FUS accelerated by disease mutation. Cell. 2015;**162**(5):1066-1077

[210] Soo KY, Sultana J, King AE, Atkinson R, Warraich ST, Sundaramoorthy V, et al. ALS-associated mutant FUS inhibits macroautophagy which is restored by overexpression of Rab1. Cell Death Discovery. 2015;**1**(July):15030

[211] Van Swieten J, Spillantini MG. Hereditary frontotemporal dementia caused by Tau gene mutations. Brain Pathology. 2007;**17**(1):63-73

[212] Kuusisto E, Salminen A, Alafuzoff I. Ubiquitin-binding protein p62 is present in neuronal and glial inclusions in human tauopathies and synucleinopathies. Neuroreport. 2001;**12**(10):2085-2090

[213] Ramesh Babu J, Lamar Seibenhener M, Peng J, Strom AL, Kemppainen R, Cox N, et al. Genetic inactivation of p62 leads to accumulation of hyperphosphorylated tau and neurodegeneration. Journal of Neurochemistry. 2008;**106**(1):107-120

[214] Arendt T, Stieler JT, holzer M. Tau and tauopathies. Brain Research Bulletin. 2016;**126** (3):238-292

[215] David DC, Layfield R, Serpell L, Narain Y, Goedert M, Spillantini MG. Proteasomal degradation of tau protein. Journal of Neurochemistry. 2002;**83**(1):176-185

[216] Hatakeyama S, Matsumoto M, Kamura T, Murayama M, Chui DH, Planel E, et al. U-box protein carboxyl terminus of Hsc70-interacting protein (CHIP) mediates poly-ubiquitylation preferentially on four-repeat Tau and is involved in neurodegeneration of tauopathy. Journal of Neurochemistry. 2004;**91**(2):299-307

[217] Dolan PJ, Johnson GVW. A caspase cleaved form of tau is preferentially degraded through the autophagy pathway. The Journal of Biological Chemistry. 2010;**285**(29):21978-21987

[218] Hamano T, Gendron TF, Causevic E, Yen SH, Lin WL, Isidoro C, et al. Autophagic-lysosomal perturbation enhances tau aggregation in transfectants with induced wild-type tau expression. The European Journal of Neuroscience. 2008;**27**(5):1119-1130

[219] Berger Z, Ravikumar B, Menzies FM, Oroz LG, Underwood BR, Pangalos MN, et al. Rapamycin alleviates toxicity of different aggregate-prone proteins. Human Molecular Genetics. 2006;**15**(3):433-442

[220] Krüger U, Wang Y, Kumar S, Mandelkow EM. Autophagic degradation of tau in primary neurons and its enhancement by trehalose. Neurobiology of Aging. 2012;**33**(10):2291-2305

[221] Schaeffer V, Goedert M. Stimulation of autophagy is neuroprotective in a mouse model of human tauopathy. Autophagy. 2012;**8**(11):1686-1687

[222] Polito VA, Li H, Martini-Stoica H, Wang B, Yang L, Xu Y, et al. Selective clearance of aberrant tau proteins and rescue of neurotoxicity by transcription factor EB. EMBO Molecular Medicine. 2014;**6**(9):1142-1160

[223] Witman GB, Cleveland DW, Weingarten MD, Kirschner MW. Tubulin requires tau for growth onto microtubule initiating sites (flagella/in vitro assembly/electron microscopy). Cell Biology. 1976;**73**(11):4070-4074

[224] Weingarten MD, Lockwood AH, Hwo SY, Kirschner MW. A protein factor essential for microtubule assembly. Proceedings of the National Academy of Sciences of the United States of America. 1975;**72**(5):1858-1862

[225] Adams SJ, Crook RJP, Deture M, Randle SJ, Innes AE, Yu XZ, et al. Overexpression of wild-type murine tau results in progressive tauopathy and neurodegeneration. American Journal of Pathology. 2009;**175**(4):1598-1609

[226] Santacruz K, Lewis J, Spires T, Paulson J, Kotilinek L, Ingelsson M, et al. Tau suppression in a neurodegenerative mouse model improves memory function. Science. 2005;**309**(5733):476-481

[227] Love S, Bridges LR, Case CP. Neurofibrillary tangles in Niemann-Pick disease type C. Brain. 1995;**118**:119-129

[228] Auer I, Schmidt M, Lee V, et al. Paired-Helical-Filament-Tau (PHFTAU) IN Niemann-Pick Type-C disease is similar to phftau in Alzheimers-Disease. Acta Neuropathologica. 1995;**90**:547-551

[229] Perlson E, Maday S, Fu M meng, Moughamian AJ, Holzbaur ELF. Retrograde axonal transport: Pathways to cell death? Trends in Neurosciences. 2010;**33**(7):335-344

Chapter 5

Lysosomal Degradation of Junctional Proteins

Catalina Flores-Maldonado, Odette Verdejo-Torres, Jessica Campos-Blázquez, Agustín Ruiz Cabrera, Vicky García-Hernández, Ruth Rincón-Heredia and Rubén G. Contreras

Additional information is available at the end of the chapter

http://dx.doi.org/10.5772/intechopen.69370

Abstract

Epithelial cells develop tight junctions (TJs) and cell polarity. Both properties are sensitive to environmental signals such as the epidermal growth factor (EGF) and the cardiotonic steroid ouabain. EGF is regarded as the main protector against injuries in epithelia, and ouabain is a hormone that regulates blood pressure, natriuresis, cell survival, and cell adhesion. After treatment with epidermal growth factor or ouabain, epithelial dog kidney MDCK cells undergo a drastic remodeling that includes changes in the transcription, translation, localization, and degradation of cell junction proteins. Degradation of these proteins involves selective and nonselective autophagy as well as endocytic lysosomal and proteasomal routes. The remodeling mechanism of tight junction's proteins includes the activation of Src and ERK1/ERK2 kinases, the phosphorylation and translocation into the nucleus of the transcription factor STAT3, the activation of PKC to induce the endocytosis of claudin-2, and the delivery of this protein to the lysosomes. Whole communicating junctions and desmosomes are internalized by one cell and sent to degradation by nonselective autophagy. Nonselective and selective autophagies in epithelial cells are very context dependent; nevertheless, it is clear that, together with endocytic lysosomal and proteasomal degradation, they play a key role in the remodeling and functioning of cell junctions.

Keywords: claudin-2, ouabain, EGF, MDCK, tight junctions

1. Introduction

When multicellular organisms left the sea of constant chemical composition to conquer the land, it became necessary to develop mechanisms to maintain a constant internal milieu similar to the sea that was left behind. *Epithelia* were fundamental in terrestrial conquest because these tissues avoid the loss of water, transport vectorially substances to introduce nutrients into the body, eliminate waste substances, and protect the individual from pathogens and their toxins [1]. These functions of epithelia depend on two basic characteristics of the cells that constitute them; the first one is *plasma membrane polarity*, which consists on the division of the plasma membrane in an apical, a lateral, and a basal domains, each one with characteristic structure, composition, and function [2]. Due to its continuity, the lateral and the basal domains are often referred as the basolateral domain. The second characteristic of epithelia is *cellular junctions* that bind epithelial cells to each other and to the substrate (**Figure 1**) [1, 3]. Adhesion between neighboring cells depends on the *junctional complex*, which is a group of intercellular junctions that includes the tight junctions (TJs), adherens junctions (AJs), desmosomes (Des), and communicating or GAP junctions (GAPJs) [4]. Attachment to the substrate requires the formation of two additional cellular junctions: focal adhesions (FAs) and hemidesmosomes (HDes) (**Figure 1**) [5, 6]. *Cellular junctions* and *plasma*

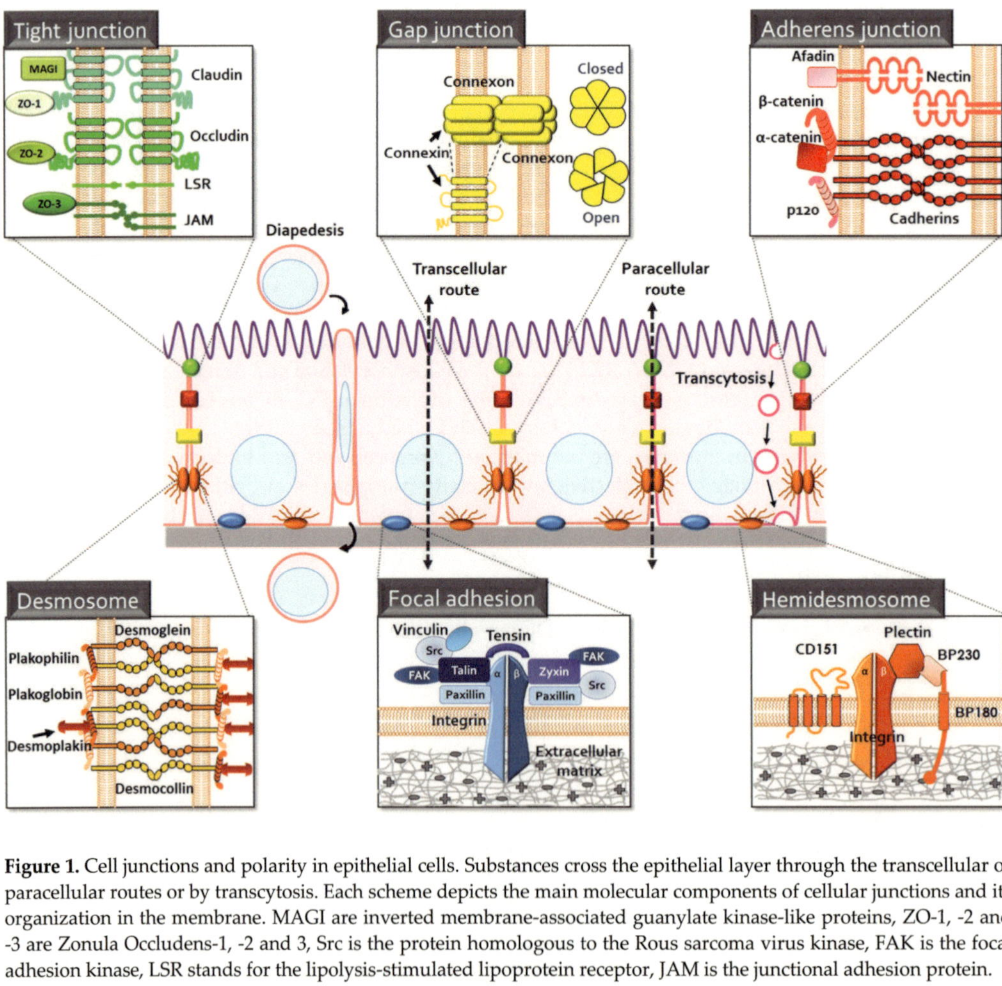

Figure 1. Cell junctions and polarity in epithelial cells. Substances cross the epithelial layer through the transcellular or paracellular routes or by transcytosis. Each scheme depicts the main molecular components of cellular junctions and its organization in the membrane. MAGI are inverted membrane-associated guanylate kinase-like proteins, ZO-1, -2 and -3 are Zonula Occludens-1, -2 and 3, Src is the protein homologous to the Rous sarcoma virus kinase, FAK is the focal adhesion kinase, LSR stands for the lipolysis-stimulated lipoprotein receptor, JAM is the junctional adhesion protein.

membrane polarity are crucial requirements for epithelial function and influence each other in complex ways, for example, the formation of a specific cell junction is the result of the polarized expression of many proteins and lipids; conversely, the development of cell polarity requires cell adhesion, initially to the extracellular matrix and then to the neighboring cells [7–10], and the development of AJs is, in general, a prerequisite for the formation of the other intercellular junctions [11, 12].

The complex organization and regulation of cell junctions and cell polarity in epithelial cells are adaptations to perform vectorial transport. A given substance crosses epithelial layers either through the *transcellular pathway*, using polarized channels, transporters, and pumps [13, 14], the *paracellular pathway*, constituted by TJs and the intercellular space [15], or both (**Figure 1**). The contribution of each pathway depends on the kind of substance transported, the cell type that executes the transport, and the physiological state of the transporting cell. In bowels and kidneys of mammalians, the transcelllular pathway accounts for most of the glucose transport [13], while flying vertebrates, perhaps because they have a smaller intestine than other vertebrates, relay in a majoritarian paracellular transport [16]; most Ca^{2+} is reabsorbed at the thick ascending limb of Henle's loop through the paracellular pathway by a specific type of claudin (CLDN), claudin-16, a TJ protein that forms Ca^{2+}- and Mg^{2+}-selective pores [17], while glucose is transported predominantly through the transcellular pathway by polarized SGTL and GLUT transporters driven by the Na^{+} concentration gradient [13, 18]. There are also two more processes to cross epithelia and endothelia: *whole cells cross epithelia*, as shown in spermatogenesis in which immature sperm cells open its way through the junctional complex of epithelial Sertoli cell monolayers [19] and in *diapedesis*, where neutrophils and leucocytes open transitorily the junctional complex and accede to the internal media [20]. Moreover, some substances and even viruses cross epithelia by *transcytosis* (**Figure 1**), a process that consists in the incorporation of the material to a vesicle generated in one plasma membrane domain that is, later on, translocated and fused to the opposite membrane domain where it delivers its content [21].

Recent research demonstrated that *cell junctions are signaling stations* that inform the adhesion status of the cells to cytoplasmic and nuclear mechanisms so they can induce proper responses [22, 23]. *Contact inhibition* and *healing of a wound* illustrate this property of cell junctions: loose AJs, or their absence by the lack of E-cadherin, an essential transmembrane protein of this cellular junction, trigger a period of high cell proliferation and migration through the activation of the Hippo signaling pathway and stop only when the normal size of the organ and a strong adhesion between epithelial cells are reached, a process referred to as *contact inhibition* [24]; when an epithelium is wounded, the cells at the front of the wound lose their cell junctions, start to proliferate, and migrate to heal the wound and extracellular protective factors, such as the epidermal growth factor (EGF), ouabain (OUA), and marinobufagenin, favor this reparation [25, 26]. Several membrane-associated proteins of cell junctions, notably β-catenin [27], a protein of the AJs, and zonula occludens-2 (ZO-2), a protein of TJs [28], are transcriptional cofactors that regulate the expression of proliferation-related genes, for instance, cyclin D1 and myc; conversely, transcription factors such as Jun, Fos, and ZONAB reside transitorily at the cytoplasmic region of the TJs [29, 30] and, at least ZONAB, downregulates paracellular permeability [30].

Cellular junctions and plasma membrane polarity are highly regulated. For example, a progressive conversion of renal intercalated cells of the collecting tubules from α to β type comprises the inversion of the apical H^+-ATPase and a basolateral Cl^-/HCO_3^+ exchanger polarity, in response to the increase in the expression of hensin, a protein of the extracellular matrix [31, 32], and cysts embedded in collagen displace their TJs from the vicinity of the lumen toward the proximity of the external surface [33, 34].

Cell junctions and plasma membrane polarity are crucial for the normal physiology of the organism, and its failure in several pathologies has disastrous consequences. To start with, it is common that the genetic elimination of crucial proteins, such as E-cadherin from the AJs or ZO-2, is lethal at embryonic stages [35, 36], but whenever an epithelial adhesion protein is not expressed, epithelia compartmentalization and vectorial transport are lost. For example, in hereditary familial hypomagnesemia with hypercalciuria and nephrocalcinosis, the lack of CLDN-16 and CLDN-19 impairs Ca^{2+} and Mg^{2+} reabsorption in the kidney [17, 37]; in cholestatic children's liver disease, the absence of ZO-2 and TJs provokes the invasion of bile salts into the blood [38]; in pemphigus vulgaris, the depletion of Des by autoantibodies against the desmosomal cadherin desmoglein-3 results in the formation of skin blisters [39], which can also appear if HDes are disassembled by mutations in the integrin β4, an adhesion molecule of this cell junction [40]; loss of adhesion and augmented proliferation in colon cancer are elicited by mutations that increases the cytosolic and nuclear pools of β-catenin [27]; infection and inflammation boost *diapedesis* [20]; rotaviruses that cause diarrheas open TJs and accede to their basolateral receptors by exposing its VP8 protein, which bear small peptides with sequences identical to some regions of occludin, an integral membrane protein of the TJs [41]; and the exogenous expression of the E6 oncoprotein of the high-risk papilloma virus type 16 induces the translocation of β-catenin into the nucleus to increase proliferation in the skin [42].

2. Cell junctions have a general common layout

All junctions have a similar structural layout: they have transmembrane proteins that are the receptors for adhesion, and a series of membrane-associated proteins that bind the cytoplasmic aspect of transmembrane receptors to the actin, tubulin, or cytokeratin cytoskeleton to provide mechanical strength. Besides cell adhesion, cell junctions are sensors that inform, in and out, the state of extracellular environment to modulate cell's proliferation, differentiation, and fate. Given that lysosomes are of paramount importance for cell junctions and plasma membrane polarity, it is necessary to briefly review the degradation routes where this organelle intervenes.

3. Lysosome degradation pathways

Lysosomes are major degradative organelles of eukaryotic cells. They were first identified as cell compartments enriched in hydrolases [43], but now they are also recognized as providers of building blocks during starvation and powerful stations to sense nutrients and regulate transcription and cellular homeostasis [44]. Lysosomes have a highly acid lumen (pH 4.5–5.0) produced by a vacuolar H^+-ATPase. The acidic pH is necessary for the hydrolysis of waste materials and drives the transport of sugars, amino acids, nucleotides, and lipids, through the single membrane of the organelle for recycling [45]. The lysosomal membrane owes its resistance against the activity of the hydrolases that it contains, to the expression of a prominent glycocalyx in its inner surface, formed by glycosylated transmembrane proteins such as the human LIMP-2 and its homologues in *Caenorhabditis elegans* SCAV-3 [46].

Many intracellular proteins are ubiquitiated and degraded in the proteasome (**Figure 2, 1**). There are also several routes to deliver cellular material into the lysosomes: an *endosomal-lysosomal* route, similar to the secretory route, that consists in the ubiquitin (Ub) or clathrin mediated endocytosis of the protein to be degraded, the posterior fusion of the vesicles containing this protein to the early endosomes (EE), which in turn may mature to late endosome (LE, 2) or produce multivesicular bodies (MVB, 3) and fuse with the lysosome (**Figures 2** and **3**) [47, 48]; the *microautophagy* degrades intracellular proteins engulfed by the lysosome (**Figure 2, 4**) [49], the nonselective *macroautophagy* degrades big intracellular waste (**Figure 2, 6**) [52, 53] (**Figure 2, 5**). *Endocytosis* requires the ESCRT machinery and the activity of several Rab GTPases that tether and dock the SNARE fusion machinery [54]. Rab5 participates in the fusion of clathrin-coated vesicles with early endosomes (**Figure 2**, EE) [55, 56], Rab4 in the recycling of proteins to the plasma membrane [57], and Rab7 in the transport from early to late endosome and lysosome biogenesis and maintenance [58]. *Nonselective macroautophagy* consists of the engulfment of intracellular waste material by a phagophore, a double-membrane structure possibly derived from the endoplasmic reticulum [53]. Since its appearance and through all its maturation, the phagophore incorporates the microtubule-associated protein 1A/1B light chain 3 (LC3). LC3 exist in a cytosolic form (LC3I) that, at the onset of

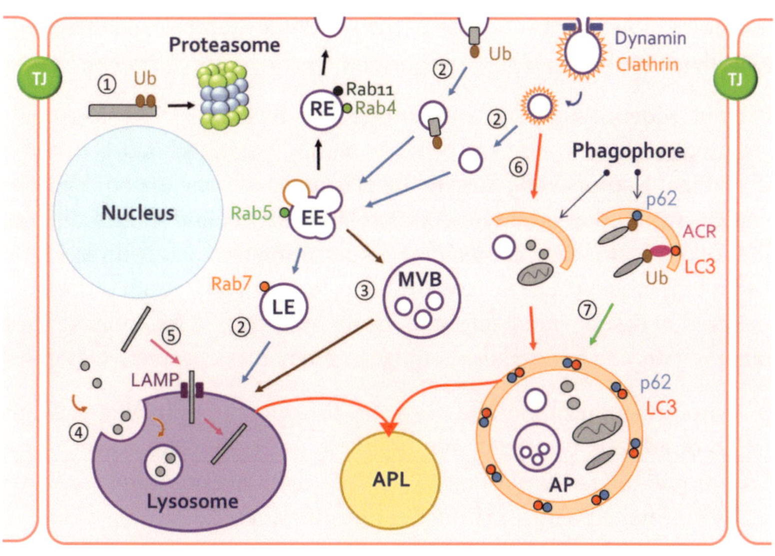

Figure 2. Protein Degradation pathways. (1) *Proteasome pathway*, used for degradation of the ubiquitin (Ub) bound cytosolic proteins. (2) Endocytic pathway, involving endocytic vesicles that fuse to the early endosome (EE), progress to the late endosome (LE), multivesicular bodies (MVB, (3) and finally to the lysosome. (4) *Microautophagy*, the lysosome engulfes some cytosolic proteins. (5) *Chaperon mediated autophagy*, allows the passage of cytosolic proteins via translocons. (6) *Nonselective macro autophagy*, large intracellular waste material is engulfed by the phagophores that becomes, later on, the autophagosome (AP). This organelle then fuses with the lysosome to form the autophagolysosome (APL). (7) *Selective macro autophagy*, intracellular waste material is ubiquitinated, bound to autophagic cargo receptors (ACR), LC3 or p62, and then engulfed by the phagophores to continue the autophagy route. RE is the recycling endosome.

autophagy, is conjugated with the lipid phosphatidylethanolamine to form LC3 type II (LC3II). *Selective macroautophagy* is triggered with the purpose to degrade a specific protein. This protein is first ubiquitinated by ligases such as Beclin1, which is then bound to the ubiquitin-binding sequestosome 1 protein (p62), which in turn binds to LC3II. The protein can also be bound, indirectly, to p62 through the autophagic cargo receptor protein (ACR, **Figure 2**, 7). Progressive encirclement of the waste material by the developing phagophore conduces to the formation of the organelle called autophagosome, which later on fuses with the lysosome to become an autophagolysosome. Fusion of the autophagosome to the lysosome requires a low acidic pH of the lysosome and is inhibited by drugs that access the lumen of this organelle and neutralize its pH, such as chloroquine and NH_4Cl [51], or inhibit the lysosomal V type H^+-ATPase, as bafilomycin A1 [59]. The lysosome itself may invaginate small cytoplasmic waste materials, in a process similar to the formation of multivesicular bodies from the late endosome [60]. In the chaperone-mediated autophagy, cytosolic unfolded proteins that contain a KFERQ-like pentapeptide bind to the lysosome receptor Lamp-2 and are directly translocated into the lumen of the lysosome through a multimeric translocation complex [50, 53].

A central regulator of lysosomal activity, particularly autophagy, is the target of rapamycin (mTOR), a multi-protein complex that includes the kinase mTOR itself, inhibited by rapamycin, the raptor adaptor, two intrinsic inhibitors of mTOR activity, DEPTOR and PRAS40, and a G-protein. The mTOR complex senses energy and nutrient availability, growth factors, and stress conditions to modify cell growth and proliferation. In normal conditions, mTOR localizes in the cytosol and triggers anabolic programs, like mRNA translation. Under starvation, mTOR is translocated to the cytosolic side of the lysosome membrane, where it initiates catabolic processes like autophagy [44].

A growing body of evidence suggests that lysosomes can function as Ca^{2+} stores and contact intimately to the endoplasmic reticulum, the peroxisome, and the mitochondria to deliver necessary lipids [61, 62] and that lysosomes can fuse to the plasma membrane to pour hydrolytic enzymes in the extracellular media that modify the extracellular matrix and induce differentiation [59]. Lysosomes induce cell death when its membrane is permeabilized and hydrolases such as cathepsin B, a Ca^{2+}-sensitive protease, are released in the cytoplasm. Cell death induced by lysosomal damage is observed in tissue remodeling, elimination of excessive intracellular waste or metals, and the immune response to intracellular pathogens and neurodegenerative diseases [63, 64].

To maintain compartmentalization and vectorial transport in epithelial cells, the synthesis and degradation of adhesion proteins must be closely coordinated. Nevertheless, epithelial cells must have certain degree of plasticity to modify cell junctions in response to the variable environment. Lysosomal activity is crucial in both situations.

4. Role of lysosomes in the degradation of tight junction proteins

4.1. Biology of tight junctions

TJs, also known as *zonula occludens* (ZO), are formed at the most apical region of the lateral plasma membrane domain [4], constitute the frontier between the apical and the basolateral domains, and obliterate the intercellular space at this region, converting the epithelia in an effective barrier against the free diffusion of ions and other substances [3]. This cellular junction has two main functions: it is a *gate* that transports ions and other substances through the paracellular pathway [65–67] and a *fence* that impairs the diffusion of integral membrane proteins and lipids from the apical to the basolateral domain and the other way around [68, 69], a property that helps to maintain cell polarity [8]. In the case of lipids, its impairment is restricted to those from the exoplasmic leaflet of the plasma membrane. In some experimental conditions, TJs can be a *bridge* that allows the diffusion, in the plane of the membrane, of membrane liposoluble probes and of phosphatidylinositol bisphosphate (PIP2), from one epithelial cell to its neighbors [70–74]. TJs are composed of three main types of integral proteins that perform the task of adhesion receptors: the four-transmembrane-domain family of CLDNs [75] and TAMP proteins, that includes occludin and tricellulin [76], the single transmembrane domain proteins of the junctional adhesion molecule (JAM) family and the lipolysis-stimulated lipoprotein receptor (LSR) (**Figure 1**) [77–82]. CLDNs constitute the paracellular ion and water channels and barriers [17, 65, 83] and confer to the TJs and the epithelia its specific paracellular permeability and selectivity, depending on the type of CLDNs expressed [17, 65, 83–85]. The integral membrane proteins of the TJs bind to a cluster of three cytoplasmic proteins of the ZO, named ZO-1, ZO-2, and ZO-3 (**Figure 1**, green). These proteins belong to the MAGUK family and are platforms for protein-protein interaction [86–88]. ZO's plaque proteins in turn bind transmembrane once to the actin cytoskeleton [89]. There are a number of other integral and associated proteins that are nonessential or reside transitorily at TJs but, nevertheless, are important for vesicular trafficking and signaling [78]. In insects and mollusks, the paracellular pathway is controlled by the septate junctions, structures that have different molecular compositions and structures and are located bellow the AJs [90]. Based on the crystal structure of the euglena tetraspan protein IP39, another member of the group of proteins to which CLDNs belong, the PMP22/EMP/MP20/Claudin superfamily, it has been suggested that CLDNs may be originated from an ancestral four-transmembrane-domain protein, similar to IP39, which may had played a role in the peculiar tortuous movement of this photosynthetic protist [91].

Epithelia adjust the permeability of their paracellular route in response to physiological requirements, pathological conditions, and pharmacological challenges. One simple way to gaze epithelial permeability is to measure the transepithelial electrical resistance (TER) [92]: the higher the value of TER, the lower the paracellular permeability. On this regard, the renal system is very illustrative. Human kidneys filtrate 170 l of plasma but secrete only 1.7 l of urine. Water, proteins, sugars, and ions from the glomerular filtrate are reabsorbed, and the filtrate is steadily concentrated along the nephron. The epithelia that line this tubular surface in vertebrate species gradually increase their TER from approximately 10 Ω cm^2 at the proximal convoluted tubule [93, 94] to several thousands of Ω cm^2 at the collecting duct [95, 96] and up to hundreds of thousands of Ω cm^2 at the bladder [97, 98]. A number of epithelial cell adaptations account for by this TER gradient: increments in cell size, reduction of the junctional membrane tortuosity, a progressive increase in the structural complexity of TJ strands, and the expression of a specific set of CLDNs in each nephron segment [15, 65]. CLDN-2 induces a low TER phenotype in renal MDCK cells [99], from cation and water-selective channels [83, 100, 101], and it is expressed in proximal tubules [102–104], where it is necessary for the uptake of Na^+, water and, likely, Ca^{2+} [105]. CLDN-4 induces a high-resistance phenotype upon the epithelial cells that express it [106–108], including those at the distal nephron segment epithelium [102–104].

4.2. Autophagy of CLDN-2 participates in the differentiation of epithelia induced by the epidermal growth factor and exerts a protective effect

The fluids that bathe apical membranes, such as urine, semen, and milk, are radically different from each other, but the interstitial milieu that contacts the basolateral membranes has a constant composition maintained by homeostatic mechanisms. This difference suggests that substances in the apical media might regulate specific epithelial properties. Several substances in the extracellular milieu induce TER changes in canine MDCK cells [109]. One of them is EGF [110], a substance previously known to increase the TER of epithelial kidney pig LL-CPK1 cells [111]. Urinary EGF reduces the cellular CLDN-1 and CLDN-2 protein level and increases CLDN-4 one [110]. EGF decreases the cellular level of CLDN-2 [112] through the simultaneous activation of Src kinase, extracellular regulated kinases 1/2 (ERK1/2) [113, 114], and the transcription factor STAT3 [114] that, in turn, may accelerate clathrin-mediated endocytosis and lysosomal degradation of CLDN-2 [113], block CLDN-2 [115], and trigger CLDN-4 [116] transcription in MDCK cells. In lung cancer cells though, EGF increases CLDN-2 through the activation of the EGF/EGFR/MEK and cFos pathway [117]. It would be interesting to find out the molecular mechanisms that fail in cancer and provoke the opposite response.

The response elicited by EGF is transient, reaches a maximal value of TER at 15 h, and slowly decreases to control values at 24 h. This downregulation is provoked by the induction of the synthesis of prostaglandin E_2 by the EGF itself that increases AmpC production, which in turn blocks the activation of ERK1/2 [118].

It is not clear which vesicular compartment participates in the CLDN-2 degradation induced by EGF. The observation that the knockdown of Rab14 induces the lysosomal degradation of CLDN-2 in MDCK cells [119] opens the possibility that EGF somehow be able to inactivate this Rab protein.

The induction of CLDN-2 downregulation by EGF is blocked by bafilomycin A1 and chloroquine, indicating that it may be performed by autophagy (**Figure 3**). A schematic representation of the mechanisms of EGF effect on CLDNs is shown in **Figure 4**. The induction of selective autophagy by EGF can be seen as a differentiation or protective effect. In this respect, autophagy has been observed in Caco-2 cancer colon cells deprived of nutrients, where selective autophagy of CLDN-2 is activated, resulting in an increase in TER [120]. Moreover, in porcine gut IPEC-1 epithelial cells, the deprivation of nonessential amino acids induces an apoptotic process that degrades CLDN-1 and ZO-1, but if autophagy is inhibited with 3MA, degradation of adhesion proteins and apoptosis is potentiated, indicating that autophagy has a protective role in these cells [121]. Finally, the injured spinal cord in rats induces the degradation of p120 and β-catenins, as well as CLDN-5 and occludin, in blood vessels of endothelia. This degradation is performed through selective autophagy, considering that these proteins associate to LC3II and p62. The administration of retinoic acid potencies autophagy and improves movement of the injured rats [122].

4.3. Ouabain induces degradation of CLDN-2

In the eighteenth century, William Withering used extracts of the herb foxglove (*Digitalis purpurea*) to treat successfully what is now called congestive heart failure [123]. The active

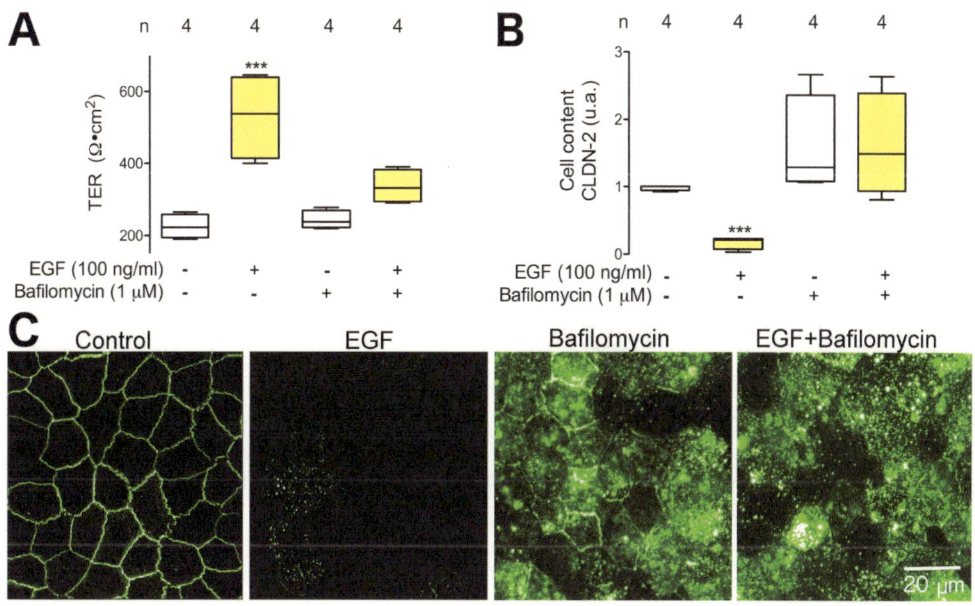

Figure 3. Epidermal growth factor (EGF) induces CLDN-2 degradation in a bafilomycin 1A-sensitive manner. Epithelial dog kidney cells (MDCK) confluent monolayer grown on filters were incubated 15 h in control condition, EGF, bafilomycin A1 (Baf), or EGF plus Baf. (A) Transepithelial electrical resistance measurements. (B) Densitometric analysis of the cellular content of CLDN-2 measured by immunoblot. (C) CLDN-2 Immunofluorescence of cells incubated in the indicated conditions.

principle involved in this therapy is digoxin, a member of the family of compounds named cardiotonic steroids, which bind and inhibit the Na^+,K^+-ATPase [124] and of which *ouabain* is the prototypic compound. This enzyme is expressed in the plasma membrane of most animal cells, where it transports intracellular Na^+ out of the cells, in exchange of extracellular K^+ toward the cell interior, against their concentration gradients, in a reaction driven by ATP hydrolysis. The inhibition by OUA became the main criterion in determining whether a given flux depended upon an active transport. Animals, including humans, produce OUA and other similar substances such as marinobufagenin and proscillaridin that are synthesized in the hypothalamus [125] and the adrenal glands [126]. Cardiotonic steroids are now considered hormones [127, 128] that regulate salt-sensitive blood pressure [129], salt handling in the kidney [130], and sodium homeostasis [131]. OUA effects require binding to the Na^+,K^+-ATPase and the activation of intracellular pathways and genes [132], including ERK1/2 [133] and the inositol (3,4,5)-tris-phosphate receptor [134, 135], indicating that Na^+,K^+-ATPase is a receptor that transduces cardiotonic steroid occupancy into intracellular mechanism. Interestingly, signaling proceeds even when transport activity is completely inhibited [136–138].

The activation of the Src-EGFR-ERK1/2 cascade by OUA regulates cell adhesion in a concentration-dependent manner: 10 nM OUA, a concentration near the hormonal level, increases the degree of sealing of the TJs, inducing the transcription, translation, and expression at the TJs of CLDNs [139], and 300 nM or higher concentrations of OUA promote cell detachment resulting from TJ, AJ, De, GAPJ, and FA disassembly, endocytosis,

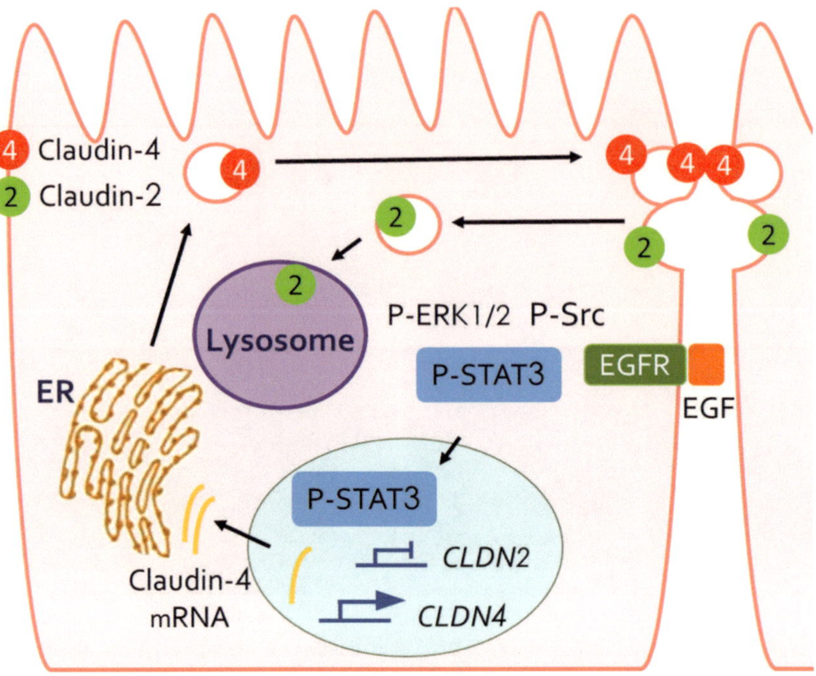

Figure 4. EGF triggers a Src-ERK1/ERK2-STAT3 cascade to induce the degradation of CLDN-2 in the lysosomes. Occupancy of EGFR by its ligand induces the simultaneous phosphorylation of the kinases Src, ERK1/ERK2, as well as the phosphorylation of the transcription factor STAT3. The phosphorylated STAT3 is translocated into the nucleus, where it upregulates the transcription of CLDN-4 at the same time downregulates that of CLDN-2. In the cytoplasm, the same cascade plays a role in the induction of CLDN-2 endocytosis and CLDN-4 exocytic fusion, events that result in the lysosomal degradation of CLDN-2, an insertion of CLDN-4 at the TJs, and an increment of the degree of sealing of the TJs.

and posterior degradation of their cell adhesion molecules [140]. Occludin, CLDN-2, and CLDN-4 endocytoses are clathrin-dependent [141]. 300 nM but not 10 nM OUA increases p62 signal and its colocalization with CLDN-2 in MDCK cells; degradation of CLDNs at 300 nM OUA is inhibited with NH_4Cl and bafilomycin A1 [141], suggesting that ouabain activates CLDN-2 degradation through autophagy (**Figure 5**). OUA increments the size of intracellular structures that bind an antibody against Rab11, a recycling endosome marker, indicating that OUA is not inducing recycling of CLDN-2 (**Figure 6**). The mechanism of OUA action is shown in **Figure 7**.

The final outcome of the treatment with high concentrations of OUA is the detachment and death of OUA-sensitive cells [132, 137, 140, 142]. Ionic imbalance that results from the inhibition of the enzyme has been considered the prime cause of cell death given the fame of the Na^+,K^+-ATPase as an ion transporter. However, cells do not detach when they are cultured in low K^+ medium [137], which mimics the diminished $[K+]_i$ content induced by OUA. The cytotoxic action of OUA in humans and rodent cells depends on the features of the type α subunit expressed, rather than by any downstream components of the cell

http://dx.doi.org/10.5772/intechopen.69370

death machinery [142]. In this respect, epithelial cells expressing a OUA-resistant isoform of Na^+,K^+-ATPase do not detach when they are incubated in low K^+- or K^+-free medium [137, 143]. Therefore, ionic imbalance by itself is not sufficient to detach cells; OUA and the activation of kinases (p38 tyrosine kinases, Src, and ERK1/2) are necessary (**Figure 7**), a finding that agrees with the triple role of the Na^+,K^+-ATPase: transporter, signaling receptor, and cell-cell adhesion molecule [144].

4.4. Autophagy plays a role in claudin-16 degradation in lysosomes

Renal hypomagnesemia with hypercalciuria and nephrocalcinosis is an autosomal recessive disease characterized by abundant renal Mg^{2+} and Ca^{2+} wasting that causes renal

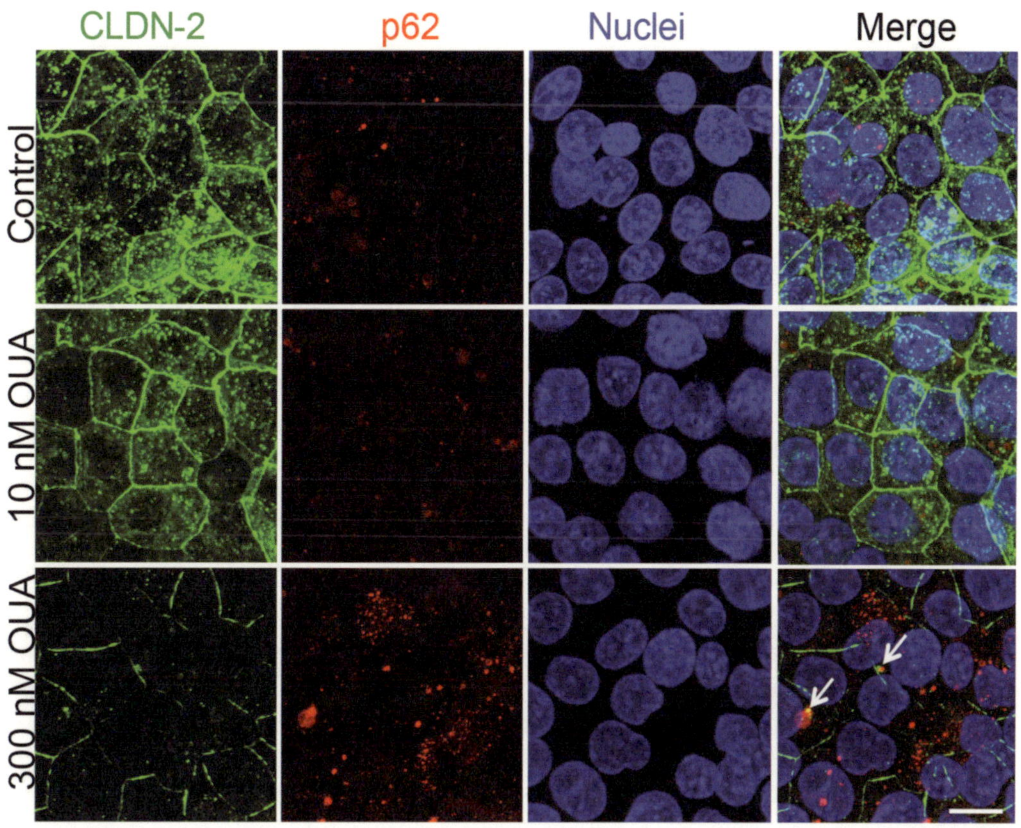

Figure 5. High ouabain concentrations increase autophagy. Control MDCK cells have their CLDN-2 localized at the TJs, in a normal quantity, and in the cytoplasm in numerous spots (green); p62 shows no colocalization with CLDN-2. Upon incubation in media containing ouabain (OUA) 10 nM images remains unchanged, indicating that low OUA concentration does not activate autophagy. On the contrary, the incubation with OUA 300 nM decreases CLDN-2 all around the cell and increases p62 that colocalizes with internalized CLDN-2 (white arrows). This result supports the observation that 300 nM OUA increases autophagy of CLDN-2. Confluent monolayers of MDCK cells were grown on coverslips overnight and then incubated with control media, ouabain 10 nM or 300 nM for 20 h. Barr corresponds to 10 μm.

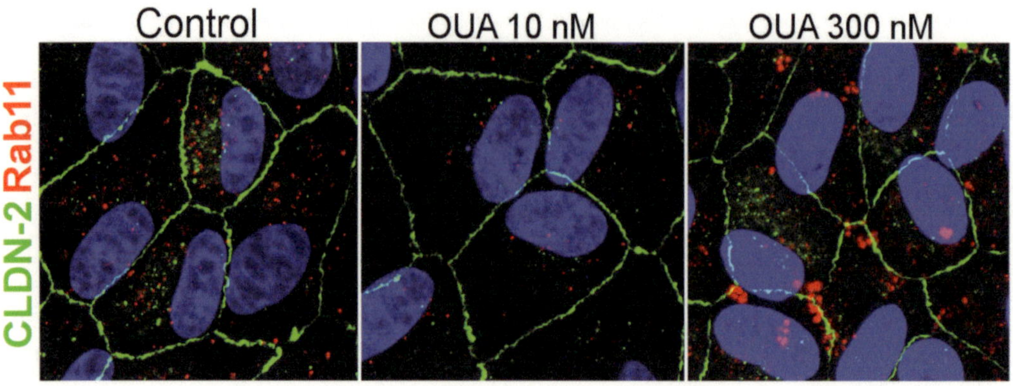

Figure 6. Ouabain does not induce recycling of CLDN-2. MDCK cells monoloyers were plated on glass coverslips overnight and incubated in control conditions (control) or in media with OUA 10 or 300 nM for 6 h. Cells were stained with antibodies against CLDN-2 and Rab11, a small GTP-binding protein of the recycling endosome. At this time, CDLN-2 has not been degraded yet and does not colocalize with Rab11 at any condition, suggesting that CLDN-2 is not internalized through the recycling endosome. Nevertheless, ouabain 10 nM decreases the intensity of the signal and the number of spots observed in the cytoplasm, implying that low OUA concentrations decrease recycling, while 300 nM increases the signal as it corresponds to cells with very active endocytosis. These results indicate that under OUA stimuli there seems to be a very active endocytic pathway, but CLDN-2 is not being recycled nor internalized through it.

parenchymal calcification and renal failure. It can only be cured through renal transplantation. The illness results of the lack of stable expression of CLDN-16 and/or CLDN-19 caused by mutations in *CLDN-16* and *CLDN-19* genes [145, 146]. Patients with a mutation in CDLN-19 show hypomagnesemia, yet this symptom is accompanied by malfunction of several glands, as well as visual impairment [146, 147]. Claudin-16 protein forms paracellular Ca^{2+} pores, while CLDN-19 forms anion impermeable TJs [148, 149]. Both CLDNs form stable membrane dimers able to conduce cations which are bound to the dimers of the neighboring cell. Dimer formation depends on the lateral interaction of their third and fourth transmembrane domains of both CLDNs. Mutations in these domains, which impair dimerization, decrease transepithelial permeability [150]. Most known disease-related mutations of CLDN-16 provoke protein misfolding and induce the accumulation of the defective protein in the endoplasmic reticulum and its degradation in the proteasome [151, 152]; other CLDN-16 mutants reach the plasma membrane but are unable to bind the ZO-1 scaffold; consequently, they are efficiently endocyted, delivered to the lysosome, and degraded there [151, 153]. Inhibitors of endocytosis may provide novel therapeutic strategies [152]. CLDN-16 phosphorylation of serine 217 stabilizes it in TJ, but when this phosphorylation is inhibited, CLDN-16 is dissociated from ZO-1 and sent to degradation in lysosome, which decreases TER and increases the transport of Mg^{2+} from the apical to the basolateral domain [154]. However, phosphorylation can also induce degradation in lysosomes or proteasomes [151]. Thus, in salivary glandular epithelial cell, SMG-C6, the activation of M3 muscarinic receptor with carbachol switches on ERK1/2, which in turn phosphorylates CLDN-4 at the serine 195. The phosphorylated CLDN-4 is endocyted in clathrin-coated vesicles and subsequently degraded in the proteasome, a phenomenon that decreases TER [155].

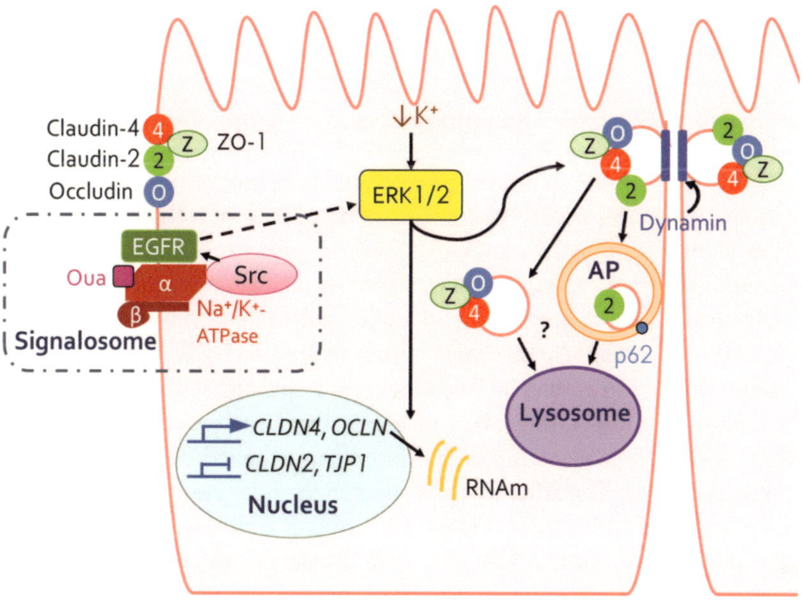

Figure 7. A high concentration of ouabain induces endocytosis and lysosomal degradation of claudins. OUA induces the formation of the signalosome (structure enclosed by the interrupted line), a caveolar complex including some Na^+,K^+-ATPases, and their associated Src and EGF receptors (EGFR). OUA activates the Src-ERK1/ERK2 pathway, which induces the clathrin- and dynamin-dependent endocytosis of TJ components. Our results indicate that there are two types of endocytic vesicles: one containing a core complex with essential TJ proteins, such as ZO-1, OCLN, and CLDN-4, and a second one consisting of components that confer a differentiated functional characteristic to TJs, such as CLDN-2, that makes TJs permeable to water and Na^+. Src-ERK1/ERK2 pathway is also required to reduce CLDN-2 and ZO-1 mRNA levels. Surprisingly, during the OUA-induced aperture of the TJs, the cellular content of CLDN-4 and OCLN mRNAs increases. *TJP1* is the gene that codes for ZO-1, and EGFR is the receptor of EGF. Ionic imbalance, indicated by the lowering of K+, may also activate ERK1/2.

5. Role of lysosomes in the degradation of adherens junctions

5.1. Biology of adherens junctions

In chordates, AJs are Ca^{2+}-dependent cell-cell adhesions between neighboring epithelial cells at the lateral domain, immediately below the TJs (**Figure 1**, red). In prechordates, AJs present an inverted localization with respect to the TJs: AJs are the most apical junction of the lateral membrane, placed over the septated junctions. AJs consist of the nectin-afadine and the cadherin-catenin complexes. The nectin's complex forms a scaffold necessary for the assembly of the AJs [156–159], whereas the cadherins serve as homotypic adhesion receptors [160, 161]. The associated plaque proteins catenins and afadins, in turn, bind the receptors to the cytoskeleton of actin [162, 163]. The homotypic adhesion of cadherin plays an important morphogenetic role because it underlays the selection and association of cells of the same type to form specific tissues, a process denominated "cell sorting" [164]. Based on the fact that there are numerous cadherins in the unicellular choanoflagellate *Monosiga brevicollis*, and that some of these proteins are expressed on the collar, a structure rich in sticking microvilla that traps and endocyte bacteria from the media to get nutrients, it has been suggested that cadherin-based intercellular adhesion may be originated from the co-option of the "trapping food" function of ancestral cadherins of unicellular organisms [165].

5.2. Endocytic route and selective autophagy degrade key proteins of adherens junctions

In a normal epithelium, β-catenin is mostly associated to E-cadherin at the plasma membrane, and the cytosolic pool of β-catenin is kept low by degradation in the proteasome. However, a proliferation signal, triggered by a Wnt ligand, impedes the β-catenin degradation and induces its accumulation in the nucleus to activate proliferation (**Figure 8B**) [27]. E-cadherin is degraded by lysosomes through an endocytic route [166]. The cytoplasmic domain of E-cadherin has an endocytosis signal that is normally masked by α, β and p120 catenins [167]. The induction of E-cadherin endocytosis and degradation starts with the tyrosine phosphorylation of E-cadherin by the kinase Src that promotes the disassembly of the cadherin-catenin complex. Then, the phosphorylated E-cadherin is ubiquitinated on the cytosolic domain by the ubiquitin ligase Hakai. The adhesion protein is then endocyted and delivered to the lysosomes via early and late endosomes (**Figure 8A**) [168]. β-Catenin may be degraded through an alternative mechanism. Thus, when there are nutrients available, a low normal level of autophagy results from the fact that most β-catenin is repressing the transcription of p62. Under starvation, β-catenin interacts with LC3II and is itself targeted for autophagic degradation (**Figure 8B**) [169]. During embryonic development, cadherin-6 restrains autophagy and drives an epithelial-mesenchymal transition (EMT) to allow a mesenchymal migratory phenotype, which is exacerbated in thyroid cancer [170]. As it happens with other junctions, the role of β-catenin is context dependent. The EMT induced by the transforming growth factor β1 in mouse kidney proximal tubular epithelial C1.1 cells is accompanied by induction of autophagy and by the nuclear translocation of β-catenin [171].

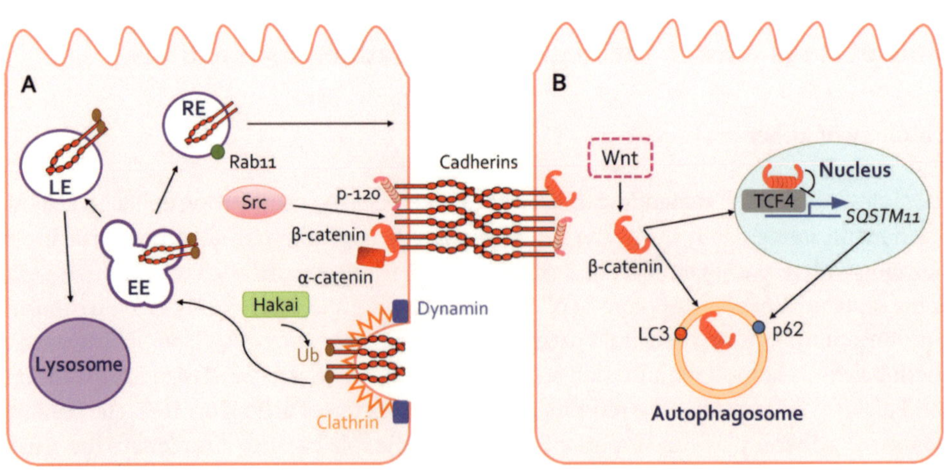

Figure 8. Degradation of E-cadherin and β-catenin involves endosomal lysosomal and autophagyc routes. (A) Normal conditions. When nutrients abound, E-cadherin and β-catenin are degraded through the endocytic-lysosomal and proteasomal routes, respectively. (B) Starvation. Under nutrient shortage, β-catenin switches to a selective macro autophagy for degradation. Wnt represents the WNT signaling cascade, TCF4 is the transcription factor 1.

http://dx.doi.org/10.5772/intechopen.69370

6. The peculiar nonselective autophagy mechanisms of degradation of large portions of communicating junctions and desmosomes

6.1. Biology of communicating junctions and desmosomes

GAPJs are molecular ducts that communicate the cytoplasm of contiguous cells and allow the epithelium to respond coordinately to various stimuli or extracellular signals (**Figure 1**, yellow). These junctions are made up of tetraspan proteins: connexins in chordates and innexins in prechordates [172]. Six connexins polymerize to form a hemichannel or connexon in a cell, which attaches to a connexon in the neighboring cell, forming in this manner an intercellular channel that can be opened by diverse stimuli. The dense clustering of tens to thousands of intercellular channels originates a GAPJ [173–175]. Connexins are associated with a scaffold of ZO-1 or ZO-3, vinculin, Src, and tubulin [176]. This association is important for the localization of connexons, the formation of the multimolecular clusters of intercellular channels in the plasma membrane, and the regulation of intercellular communication [177].

Des are cell-to-cell adhesion structures that confer mechanical strength to epithelia and cardiomyocytes. These junctions are composed of five main proteins: the desmosomal cadherins, desmogleins, and desmocollins are the receptors for adhesion. Their cytoplasmic tails bind to plaque proteins of the armadillo family, plakoglobin and plakophilin (**Figure 1**) [178]. The armadillo proteins attach to another plaque protein, desmoplakin, which, in turn, links the protein cluster to the cytoskeleton made of intermediary filaments of cytokeratin [179]. Observations in tissues and cultured cells have shown that Des can adopt a Ca^{2+}-dependent adhesion state that progresses to a Ca^{2+}-independent hyper-adhesion state, a process that requires PKC activation [22, 180–182].

6.2. Macromolecular assemblies of communicating junctions and desmosomes are degraded by autophagy

GAPJs are extremely stable junctional structures: as soon as they are formed, they become indestructible [183, 184]. Nevertheless, they are very dynamic due to the fact that connexins have a very short half-life of only 1–5 h [185]. Consequently, there is a permanent turnover that involves the closure of the intercellular conduction by several stimuli, for example, the binding of EGF to its receptor. The central portion of the GAPJ is then internalized, including the bound hemichannels and membrane of the neighboring cell, forming a peculiar structure named annular GAPJ (**Figure 9**) [186]. In some conditions, annular GAPJ may be recycled back to the plasma membrane [187] although, usually, they are degraded through autophagy; yet, the precise mechanism, the kind of autophagy involved, and the fate of the cells depend on the trigger and/or the cellular context [185, 187–189]. A mechanism that stops autophagy implicates the hijacking of components of the initiation of autophagy, for example, Atg16, by the connexins themselves. On nutrient starvation, connexins release Atg16, the blockade is lost and autophagy proceeds [190].

On liver cells of BRL 3A expressing connexin-43, cadmium inhibits GAPJ intercellular communications and induces the degradation by autophagy of connexin-43 as well as apoptosis.

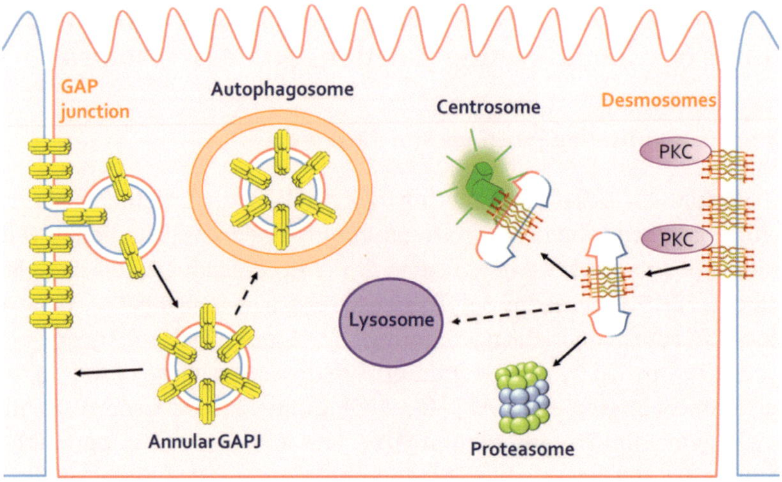

Figure 9. Big portions of GAPJ and complete desmosomes (Des) are degraded by nonselective autophagy. The central portion of the GAPJs is internalized and degraded by autophagy. Dependent on the cell type and condition, complete Des are internalized and degraded by autophagy, and halves of Des are degraded by autophagy and in the proteasome.

Inhibition of autophagy exacerbates Cd^{2+}-induced inhibition of the intercellular communication and apoptotic cell death [188] revealing the protective role that autophagy plays on cell fate.

Des are also very stable structures which can reach a hyper-adhesion state insensitive to Ca^{2+} depletion [191]. It has been shown that a half of Des is internalized after extracellular Ca^{2+} depletion in a PKC- and actin-dependent process [182, 192]. Internalized half desmosome is then transported by kinesins and microtubules toward the centrosome and remains there without recycle to the plasma membrane. Degradation proceeds in lysosomes and proteasomes [193]. In mouse epidermis, the complete Des are engulfed and internalized [192]. Nevertheless, the degradation mechanism is different when disassembly is triggered with autoantibodies from pemphigus vulgaris patients; in this case, Des disassemble in smaller complexes made of the autoantibody, desmoglein-3, and plakoglobin that are endocyted and delivered to the lysosomes through the endocytic route [194].

7. Continuous recycling in focal adhesion requires autophagy

7.1. Biology of focal adhesion and hemidesmosomes

FAs, also known as focal contacts, and HDes are the cellular junctions that attach cells to the extracellular matrix. HDes are common in stratified epithelia and bind epithelial cells to the underlying extracellular matrix (**Figure 1**, blue) [6] . The adhesion receptors of both, FAs and HDes, are transmembrane proteins of the family of integrins, which exist as heterodimers of α and β subunits form. There are 19 α-integrins and 8 β-integrins that combine to form 25 existing heterodimers in mammals [195, 196]. HDes provide stable adhesion and mechanical resistance to epithelial tissues by anchoring the extracellular matrix to the cytokeratin cytoskeleton, through a protein complex that includes the adhesion receptors α6β4 integrin, BP180, and the tetraspanin CD151, and the intracellular adapter proteins plectin and BP230 [6]. The expression of several HDes proteins depends on the transcription factor SOXF [197]. While the extracellular region of integrins of FAs binds the extracellular matrix, the cytosolic portion contacts specific plaque proteins such as focal adhesion kinase (FAK) and paxillin, which are important signaling proteins. Other protein components of the FA plaque, such as talin, vinculin, and α-actinin, bind the adhesion receptors to the actin microfilaments [5, 198, 199].

7.2. Role of lysosomes in the regulation of focal adhesions

FAs are essential in cell migration and, therefore, for embryogenesis, wound healing, immune cell function, cancer progression, and promoting metastasis [200]. Cell migration requires endocytosis and recycling of integrins given by endocytic signals in its cytoplasmic tail. These signals bind either clathrin or caveolin-1 to induce integrin endocytosis. Once inside the cell, integrins anchored to protein complexes are sent to the early endosomes, where they can be sorted either to late endosomes and lysosomes for degradation (**Figure 10**, **1**) or to recycling endosomes and plasmatic membrane for the assembly of new FAs. A short loop for recycling requires Rab4 proteins and is generally activated in response to growth factors (**Figure 10**, **2**); the long loop is Rab11 and Arf6 dependent and delivers integrins to the perinuclear recycling compartments (PNRCs) and, from there, to the cell membrane (**Figure 10**, **3**) [190]. The actin cytoskeleton is essential to the recycling pathway; in fact, depletion of the actin-related protein (Arp) 2/3 or the nucleating-promoting factors such as the members of the Wiskott-Aldrich syndrome protein (WASP) blocks recycling and induces delivery to the lysosomes [5, 198].

FA disassembly is linked to autophagy in two ways: a nonselective autophagy triggered by extreme stress condition, such as starvation or hypoxia (**Figure 10**, **4**), and a selective autophagy for housekeeping and quality control that includes ubiquitin-tagged substrate association of them with an autophagic cargo receptor (ACR) attached to LC3II. This autophagy provokes the disassemby of FA Under starvation, β1 integrin is degraded in autophagosomes in cervix adenocarcinoma epithelial HeLa cells. This autophagy is inhibited by high mTOR activity at the leading edge during migration, which promotes increased motility [201], whereas the activation of selective autophagy promotes FA disassembly in metastatic mammary epithelial cells (4T1) [202]. Thus, the inhibition of autophagosome reduces cancer cell's malignancy, indicating that selective authophagy is also a cell migration regulator (**Figure 10**).

Besides migration, autophagy is linked to anoikis, a type of cell death due to detachment from the substrate. Loss of integrin-mediated adhesion initiates autophagy, which delays anoikis and downregulates apoptotic signals. This process affords cells time to reattach; however, in cancer cells, high autophagic activity after detachment provides resistance and promotes malignancy, allowing the cell to support stress condition, increase motility, and resist anoikis [203, 204].

Although there are several illnesses produced by the lack of HDes protein expression, little is known about HDes degradation.

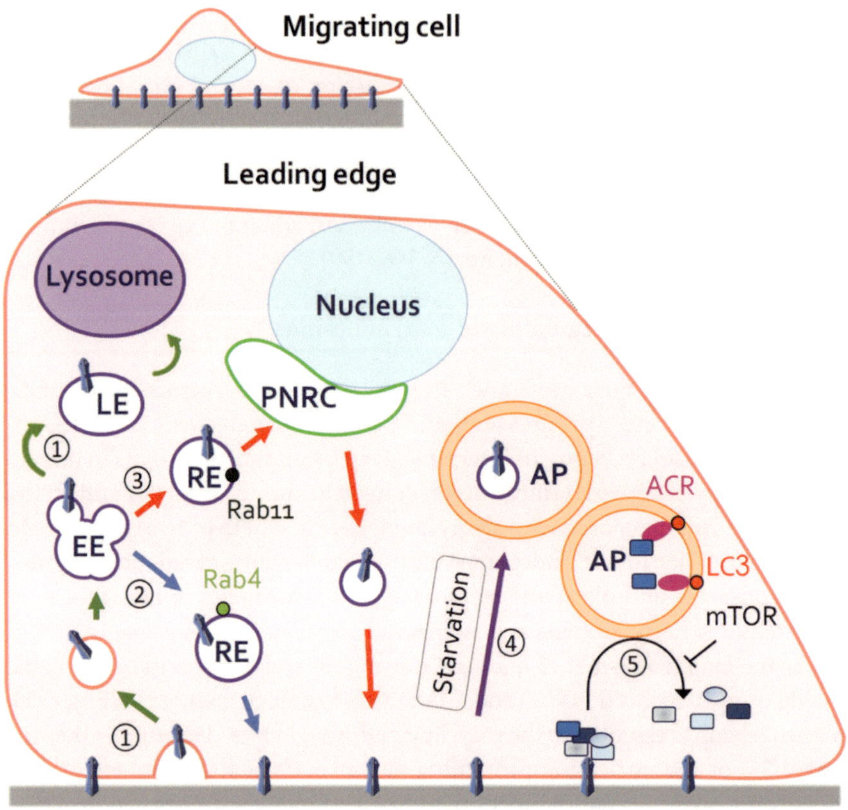

Figure 10. Autophagy is crucial for the recycling of integrins in focal adhesions during migration. Cell migration requires continuous recycling of integrins. (1) Integrin endocytic pathway degradation. (2) Integrin short loop recycling; vesicular transport of integrins from the EE to the Rab4 containing RE, and from there back to the plasma membrane. (3) Integrin long loop recycling; vesicles transport integrins from EE to Rab11 containing RE, later on, to a perinuclear recycling compartment (PNRC) and then to the plasma membrane. (4) During starvation, integrins are endocyted and directed to the AP. (5) Cell migration leading edge. FA´s protein paxillin is recognized by autophagic cargo receptors (ACR) and degraded by selective macroautophagy, which induces FAs disassembly through a mTOR dependen pathway. In the leading edge, FAs must be first formed and then degraded to allow motility. Autophagy plays a crucial role in this process.

8. Concluding remarks

Lysosomal degradation mechanisms are crucial for the formation, differentiation, and degradation of epithelial cell junctions. Epithelial cells use selective autophagy to degrade claudin-2, in response to the stimulation with the epidermal growth factor. Ouabain, at a concentration close to the hormonal, does not induce autophagy of tight junction proteins and, at high concentrations though, induces lysosomal degradation that can involve autophagy. The precise sequence of events and outcome of each lysosomal degradation mechanism is context dependent; nevertheless, it is clear that the degradation through macroautophagy of large plaque of complete communicating junctions and desmosomes, as well as of desmosomal halves, takes place either in natural tissues or in cultured cells. It is also clear that the

http://dx.doi.org/10.5772/intechopen.69370

desmosomal transition from weak to strong adhesion stages requires lysosomal activity, that β-catenin undergoes selective autophagy in some conditions and that E-cadherin degradation is performed in lysosomes through an endocytic route.

Author details

Catalina Flores-Maldonado[1], Odette Verdejo-Torres[1], Jessica Campos-Blázquez[1], Agustín Ruiz Cabrera,Vicky García-Hernández[3], Ruth Rincón-Heredia[2] and Rubén G. Contreras[1]*

*Address all correspondence to: rcontrer@fisio.cinvestav.mx

1 Departamento de Fisiología, Biofísica y Neurociencias, Cinvestav, Ciudad de México, México

2 Departamento de Neurociencia Cognitiva, Instituto de Fisiología Celular, UNAM, Ciudad de México, México

3 Department of Pathology, University of Michigan, Ann Arbor, MI, USA

References

[1] Cereijido M, Contreras RG, Shoshani L. Cell adhesion, polarity, and epithelia in the dawn of metazoans. Physiological Reviews. 2004;**84**:1229-1262. DOI: 10.1152/physrev.00001.2004

[2] Cereijido M, Contreras RG, Shoshani L, García-Villegas MR. Membrane targeting. Progress in Biophysics & Molecular Biology. 2003;**81**:81-115

[3] Cereijido M, Gonzalez-Mariscal L, Contreras RG. Epithelial tight junctions. American Review of Respiratory Disease. 1988;**138**:S17-S21. DOI: 10.1164/ajrccm/138.6_Pt_2.S17

[4] Farquhar MG, Palade GE. Junctional complexes in various epithelia. Journal of Cell Biology. 1963;**17**:375-412

[5] Burridge K, Chrzanowska-Wodnicka M. Focal adhesions, contractility, and signaling. Annual Review of Cell and Developmental Biology. 1996;**12**:463-518. DOI: 10.1146/annurev.cellbio.12.1.463

[6] de Pereda JM, Ortega E, Alonso-García N, Gómez-Hernández M, Sonnenberg A. Advances and perspectives of the architecture of hemidesmosomes: Lessons from structural biology. Cell Adhesion & Migration. 2009;**3**:361-364

[7] Cereijido M, Valdés J, Shoshani L, Contreras RG. Role of tight junctions in establishing and maintaining cell polarity. Annual Review of Physiology. 1998;**60**:161-177. DOI: 10.1146/annurev.physiol.60.1.161

[8] Cereijido M, Contreras RG, González-Mariscal L. Development and alteration of polarity. Annual Review of Physiology. 1989;**51**:785-795. DOI: 10.1146/annurev.ph.51.030189.004033

[9] Shoshani L, Contreras RG. Biogenesis of epithelial polarity and the tight junctions. In: Anderson JM, Cereijido M, editors. Tight Junctions. Boston, MA: Springer Science & Business Media. 2007. pp. 165-198

[10] Yeaman C, Grindstaff KK, Nelson WJ. New perspectives on mechanisms involved in generating epithelial cell polarity. Physiological Reviews. 1999;**79**:73-98

[11] Gumbiner B, Simons K. A functional assay for proteins involved in establishing an epithelial occluding barrier: Identification of a uvomorulin-like polypeptide. Journal of Cell Biology. 1986;**102**:457-468

[12] Gumbiner B, Simons K. The role of uvomorulin in the formation of epithelial occluding junctions. Ciba Foundation Symposium. 1987;**125**:168-186

[13] Poulsen SB, Fenton RA, Rieg T. Sodium-glucose cotransport. Current Opinion in Nephrology and Hypertension. 2015;**24**:463-469. DOI: 10.1097/MNH.0000000000000152

[14] Chen L, Tuo B, Dong H. Regulation of intestinal glucose absorption by ion channels and transporters. Nutrients. 2016;**8**:43. DOI: 10.3390/nu8010043

[15] Cereijido M, Ruiz O, Gonzalez-Mariscal L, Contreras RG, Balda MS, Garcia-Villegas R. The paracellular pathway: A small version of the kidney nephron. In: Audus KL, Raub TJ, editors. Biological Barriers to Protein Delivery. Boston, MA: Springer Science & Business Media. 1994. pp. 107-125. DOI: 10.1007/978-1-4615-2898-2

[16] Price ER, Rott KH, Caviedes-Vidal E, Karasov WH. Claudin gene expression patterns do not associate with interspecific differences in paracellular nutrient absorption. Comparative Biochemistry and Physiology—Part B: Biochemistry & Molecular Biology. 2015;**191**:36-45. DOI: 10.1016/j.cbpb.2015.09.003

[17] Simon DB, Lu Y, Choate KA, Velazquez H, Al-Sabban E, Praga M, et al. Paracellin-1, a renal tight junction protein required for paracellular Mg2+ resorption. Science. 1999;**285**:103-106

[18] Koefoed-Johsen V, Ussing HH. The nature of the frog skin potential. Acta Physiologica Scandinavica. 1958;**42**:298-308. DOI: 10.1111/j.1748-1716.1958.tb01563.x.

[19] Cheng CY, Mruk DD. Cell junction dynamics in the testis: Sertoli-germ cell interactions and male contraceptive development. Physiological Reviews. 2002;**82**:825-874. DOI: 10.1152/physrev.00009.2002

[20] Brazil JC, Parkos CA. Pathobiology of neutrophil-epithelial interactions. Immunology Reviews. 2016;**273**(1):94-111. DOI: 10.1111/imr.12446

[21] Thuenauer R, Müller SK, Römer W. Pathways of protein and lipid receptor-mediated transcytosis in drug delivery. Expert Opinion on Drug Delivery. 2016;**14**(3):1-11. DOI: 10.1080/17425247.2016.1220364

[22] Broussard JA, Getsios S, Green KJ. Desmosome regulation and signaling in disease. Cell and Tissue Research. 2015;**360**:1-12. DOI: 10.1007/s00441-015-2136-5

[23] Matter K, Balda MS. Signalling to and from tight junctions. Nature Reviews Molecular Cell Biology. 2003;**4**:225-237. DOI: 10.1038/nrm1055

[24] Kim NG, Gumbiner BM. Adhesion to fibronectin regulates Hippo signaling via the FAK-Src-PI3K pathway. Journal of Cell Biology. 2015;**210**:503-515. DOI: 10.1083/jcb.201501025

[25] Scarpa E, Mayor R. Collective cell migration in development. Journal of Cell Biology. 2016;**212**:143-155. DOI: 10.1083/jcb.201508047

[26] El-Okdi N, Smaili S, Raju V, Shidyak A, Gupta S, Fedorova L, et al. Effects of cardiotonic steroids on dermal collagen synthesis and wound healing. Journal of Applied Physiology. 2008;**105**:30-36. DOI: 10.1152/japplphysiol.00119.2008

[27] Clevers H. Wnt/beta-catenin signaling in development and disease. Cell. 2006;**127**:469-480. DOI: 10.1016/j.cell.2006.10.018

[28] Huerta M, Muñoz R, Tapia R, Soto-Reyes E, Ramírez L, Recillas-Targa F, et al. Cyclin D1 is transcriptionally down-regulated by ZO-2 via an E box and the transcription factor c-Myc. Molecular Biology of the Cell. 2007;**18**:4826-4836. DOI: 10.1091/mbc.E07-02-0109

[29] Betanzos A, Huerta M, Lopez-Bayghen E, Azuara E, Amerena J, Gonzalez-Mariscal L. The tight junction protein ZO-2 associates with Jun, Fos and C/EBP transcription factors in epithelial cells. Experimental Cell Research. 2004;**292**:51-66

[30] Balda MS, Garrett MD, Matter K. The ZO-1-associated Y-box factor ZONAB regulates epithelial cell proliferation and cell density. Journal of Cell Biology. 2003;**160**:423-432. DOI: 10.1083/jcb.200210020

[31] Takito J, Hikita C, Al-Awqati Q. Hensin, a new collecting duct protein involved in the in vitro plasticity of intercalated cell polarity. Journal of Clinical Investigation. 1996;**98**:2324-2331. DOI: 10.1172/JCI119044

[32] Al-Awqati Q, Vijayakumar S, Takito J. Terminal differentiation of epithelia. Biological Chemistry. 2003;**384**:1255-1258. DOI: 10.1515/BC.2003.138

[33] Barriere H, Chambard M, Mauchamp J, Gabrion J. Polarity reversal of inside-out thyroid follicles cultured within collagen gel: An ultrastructural study. Biology of the Cell. 1986;**57**:39-51

[34] Barriere H, Chambard M, Selzner JP, Mauchamp J, Gabrion J. Polarity reversal of inside-out thyroid follicles cultured within collagen gel: Structure of the junctions assessed by freeze-fracture and lanthanum permeability. Biology of the Cell. 1988;**62**:133-144

[35] Larue L, Ohsugi M, Hirchenhain J, Kemler R. E-cadherin null mutant embryos fail to form a trophectoderm epithelium. Proceedings of the National Academy of Sciences. 1994;**91**:8263-8267

[36] Xu J, Kausalya PJ, Phua DCY, Ali SM, Hossain Z, Hunziker W. Early embryonic lethality of mice lacking ZO-2, but Not ZO-3, reveals critical and nonredundant roles for individual zonula occludens proteins in mammalian development. Molecular and Cellular Biology. 2008;**28**:1669-1678. DOI: 10.1128/MCB.00891-07

[37] Konrad M, Schaller A, Seelow D, Pandey AV, Waldegger S, Lesslauer A, et al. Mutations in the tight-junction gene claudin 19 (CLDN19) are associated with renal magnesium wasting, renal failure, and severe ocular involvement. The American Journal of Human Genetics. 2006;**79**:949-957. DOI: 10.1086/508617

[38] Sambrotta M, Strautnieks S, Papouli E, Rushton P, Clark BE, Parry DA, et al. Mutations in TJP2 cause progressive cholestatic liver disease. Nature Genetics. 2014;**46**:326-328. DOI: 10.1038/ng.2918

[39] Samuelov L, Sprecher E. Inherited desmosomal disorders. Cell and Tissue Research. 2014;**360**(3):1-19. DOI: 10.1007/s00441-014-2062-y

[40] Koster J, Kuikman I, Kreft M, Sonnenberg A. Two different mutations in the cytoplasmic domain of the integrin beta 4 subunit in nonlethal forms of epidermolysis bullosa prevent interaction of beta 4 with plectin. Journal of Investigative Dermatology. 2001;**117**:1405-1411. DOI: 10.1046/j.0022-202x.2001.01567.x

[41] Nava P, López S, Arias CF, Islas S, Gonzalez-Mariscal L. The rotavirus surface protein VP8 modulates the gate and fence function of tight junctions in epithelial cells. Journal of Cell Science. 2004;**117**:5509-5519. DOI: 10.1242/jcs.01425

[42] Bonilla-Delgado J, Bulut G, Liu X, Cortés-Malagón EM, Schlegel R, Flores-Maldonado C, et al. The E6 oncoprotein from HPV16 enhances the canonical Wnt/β-catenin pathway in skin epidermis in vivo. Molecular Cancer Research. 2012;**10**:250-258. DOI: 10.1158/1541-7786.MCR-11-0287

[43] de Duve C. Lysosomes. Bulletin De l'Academie Royale De Medecine De Belgique. 1958;**23**:608-618

[44] Lim CY, Zoncu R. The lysosome as a command-and-control center for cellular metabolism. Journal of Cell Biology. 2016;**214**:653-664. DOI: 10.1083/jcb.201607005

[45] Forgac M. Vacuolar ATPases: Rotary proton pumps in physiology and pathophysiology. Nature Reviews Molecular Cell Biology. 2007;**8**:917-929. DOI: 10.1038/nrm2272

[46] Li Y, Chen B, Zou W, Wang X, Wu Y, Zhao D, et al. The lysosomal membrane protein SCAV-3 maintains lysosome integrity and adult longevity. Journal of Cell Biology. 2016;**215**:167-185. DOI: 10.1083/jcb.201602090

[47] Schreij AMA, Fon EA, McPherson PS. Endocytic membrane trafficking and neurodegenerative disease. Cellular and Molecular Life Sciences. 2016;**73**:1529-1545. DOI: 10.1007/s00018-015-2105-x

[48] Scharaw S, Iskar M, Ori A, Boncompain G, Laketa V, Poser I, et al. The endosomal transcriptional regulator RNF11 integrates degradation and transport of EGFR. Journal of Cell Biology. 2016;**261**:543-558. DOI: 10.1083/jcb.201601090

[49] Marzella L, Ahlberg J, Glaumann H. Autophagy, heterophagy, microautophagy and crinophagy as the means for intracellular degradation. Virchows Arch B Cell Pathol Incl Mol Pathol. 1981;**36**(2-3):219-34

[50] Orenstein SJ, Cuervo AM. Chaperone-mediated autophagy: Molecular mechanisms and physiological relevance. Seminars in Cell and Developmental Biology. 2010;**21**:719-726. DOI: 10.1016/j.semcdb.2010.02.005

[51] Mizushima N, Komatsu M. Autophagy: Renovation of cells and tissues. Cell. 2011;**147**:728-741. DOI: 10.1016/j.cell.2011.10.026

[52] Rabinowitz JD, White E. Autophagy and metabolism. Science. 2010;**330**:1344-1348. DOI: 10.1126/science.1193497

[53] Singh R, Cuervo AM. Autophagy in the cellular energetic balance. Cell Metabolism. 2011;**13**:495-504. DOI: 10.1016/j.cmet.2011.04.004

[54] Schmidt MR, Haucke V. Recycling endosomes in neuronal membrane traffic. Biology of the Cell. 2007;**99**:333-342. DOI: 10.1042/BC20070007

[55] Bucci C, Parton RG, Mather IH, Stunnenberg H, Simons K, Hoflack B, et al. The small GTPase rab5 functions as a regulatory factor in the early endocytic pathway. Cell. 1992;**70**:715-728

[56] Morrison HA, Dionne H, Rusten TE, Brech A, Fisher WW, Pfeiffer BD, et al. Regulation of early endosomal entry by the Drosophila tumor suppressors Rabenosyn and Vps45. Molecular Biology of the Cell. 2008;**19**:4167-4176. DOI: 10.1091/mbc.E08-07-0716

[57] Sönnichsen B, De Renzis S, Nielsen E, Rietdorf J, Zerial M. Distinct membrane domains on endosomes in the recycling pathway visualized by multicolor imaging of Rab4. Rab5, and Rab11. Journal of Cell Biology. 2000;**149**:901-914

[58] Bucci C, Thomsen P, Nicoziani P, McCarthy J, van Deurs B. Rab7: A key to lysosome biogenesis. Molecular Biology of the Cell. 2000;**11**:467-480

[59] Yoshimori T, Yamamoto A, Moriyama Y, Futai M, Tashiro Y. Bafilomycin A1, a specific inhibitor of vacuolar-type H(+)-ATPase, inhibits acidification and protein degradation in lysosomes of cultured cells. Journal of Biological Chemistry. 1991;**266**:17707-17712

[60] Sahu R, Kaushik S, Clement CC, Cannizzo ES, Scharf B, Follenzi A, et al. Microautophagy of cytosolic proteins by late endosomes. Developmental Cell. 2011;**20**:131-139. DOI: 10.1016/j.devcel.2010.12.003

[61] Hua R, Cheng D, Coyaud É, Freeman S, Di Pietro E, Wang Y, et al. VAPs and ACBD5 tether peroxisomes to the ER for peroxisome maintenance and lipid homeostasis. Journal of Cell Biology. 2017;**16**(2):367-377. DOI: 10.1083/jcb.201608128

[62] Costello JL, Castro IG, Hacker C, Schrader TA, Metz J, Zeuschner D, et al. ACBD5 and VAPB mediate membrane associations between peroxisomes and the ER. Journal of Cell Biology. 2017;**931**:331-342. DOI: 10.1083/jcb.201607055

[63] Gómez-Sintes R, Ledesma MD, Boya P. Lysosomal cell death mechanisms in aging. Ageing Research Reviews. 2016;**32**:150-168. DOI: 10.1016/j.arr.2016.02.009

[64] Toops KA, Lakkaraju A. Let's play a game of chutes and ladders: Lysosome fusion with the epithelial plasma membrane. Communicative & Integrative Biology. 2013;**6**:e24474. DOI: 10.4161/cib.24474

[65] Claude P, Goodenough DA. Goodenough. Fracture faces of zonulae occludentes from "tight" and "leaky" epithelia. Journal of Cell Biology. 1973;**58**:390-400

[66] Diamond JM. Twenty-first Bowditch lecture. The epithelial junction: Bridge, gate, and fence. Physiologist. 1977;**20**:10-18

[67] Mandel LJ, Bacallao R, Zampighi G. Uncoupling of the molecular 'fence' and paracellular "gate" functions in epithelial tight junctions. Nature. 1993;**361**:552-555. DOI: 10.1038/361552a0

[68] Dragsten PR, Blumenthal R, Handler JS. Membrane asymmetry in epithelia: Is the tight junction a barrier to diffusion in the plasma membrane? Nature. 1981;**294**:718-722

[69] van Meer G, Simons K. The function of tight junctions in maintaining differences in lipid composition between the apical and the basolateral cell surface domains of MDCK cells. The EMBO Journal. 1986;**5**:1455-1464

[70] Cereijido M, Contreras RG, Shoshani L, Flores-Benitez D, Larre I. Tight junction and polarity interaction in the transporting epithelial phenotype. Biochimica et Biophysica Acta. 2008;**1778**:770-793. DOI: 10.1016/j.bbamem.2007.09.001

[71] Laffafian I, Hallett MB. Lipid-protein cargo transfer: A mode of direct cell-to-cell communication for lipids and their associated proteins. Journal of Cellular Physiology. 2007;**210**:336-342. DOI: 10.1002/jcp.20851

[72] Sneyd J, Wetton BT, Charles AC, Sanderson MJ. Intercellular calcium waves mediated by diffusion of inositol trisphosphate: A two-dimensional model. American Journal of Physiology. 1995;**268**:C1537-C1545

[73] Turin L, Béhé P, Plonsky I, Dunina-Barkovskaya A. Hydrophobic ion transfer between membranes of adjacent hepatocytes: A possible probe of tight junction structure. Proceedings of the National Academy of Sciences. 1991;**88**:9365-9369

[74] Grebenkämper K, Galla HJ. Translational diffusion measurements of a fluorescent phospholipid between MDCK-I cells support the lipid model of the tight junctions. Chemistry and Physics of Lipids. 1994;**71**:133-143

[75] Furuse M, Fujita K, Hiiragi T, Fujimoto K, Tsukita S. Claudin-1 and -2: Novel integral membrane proteins localizing at tight junctions with no sequence similarity to occludin. Journal of Cell Biology. 1998;**141**:1539-1550

[76] Haseloff RF, Dithmer S, Winkler L, Wolburg H, Blasig IE. Transmembrane proteins of the tight junctions at the blood-brain barrier: Structural and functional aspects. Seminars in Cell and Developmental Biology. 2015;**38**:16-25. DOI: 10.1016/j.semcdb.2014.11.004

[77] Cereijido M, Contreras RG, Flores-Benítez D, Flores-Maldonado C, Larre I, Ruiz A, et al. New diseases derived or associated with the tight junction. Archives of Medical Research. 2007;**38**:465-478. DOI: 10.1016/j.arcmed.2007.02.003

[78] González-Mariscal L, Betanzos A, Nava P, Jaramillo BE. Tight junction proteins. Progress in Biophysics & Molecular Biology. 2003;**81**:1-44

[79] Tsukita S, Furuse M, Itoh M. Multifunctional strands in tight junctions. Nature Reviews Molecular Cell Biology. 2001;**2**:285-293. DOI: 10.1038/35067088

[80] Anderson JM. Molecular structure of tight junctions and their role in epithelial transport. Physiology. 2001;**16**:126-130. DOI: 10.1007/BF01870332

[81] Cereijido M, Anderson JM. Tight Junctions. CRC Press; 2001. USA

[82] Gonzalez-Mariscal L. Tight Junctions. Boston, MA: Springer Science & Business Media; 2007. DOI: 10.1007/0-387-36673-3

[83] Weber CR, Liang GH, Wang Y, Das S, Shen L, Yu ASL, et al. Claudin-2-dependent paracellular channels are dynamically gated. eLife Sciences. 2016;**4**:e09906. DOI: 10.7554/eLife.09906

[84] Kottra G, Frömter E. Functional properties of the paracellular pathway in some leaky epithelia. Journal of Experimental Biology. 1983;**106**:217-229

[85] Günzel D, Yu ASL. Claudins and the modulation of tight junction permeability. Physiological Reviews. 2013;**93**:525-569. DOI: 10.1152/physrev.00019.2012

[86] Anderson JM, Stevenson BR, Jesaitis LA, Goodenough DA, Mooseker MS. Characterization of ZO-1, a protein component of the tight junction from mouse liver and Madin-Darby canine kidney cells. Journal of Cell Biology. 1988;**106**:1141-1149

[87] Gonzalez-Mariscal L, Betanzos A, Avila-Flores A. MAGUK proteins: Structure and role in the tight junction. Seminars in Cell and Developmental Biology. 2000;**11**:315-324. DOI: 10.1006/scdb.2000.0178

[88] Gonzalez-Mariscal L. Molecular Characterization of the Tight Junction Protein ZO-1 in MDCK Cells*1, *2. Experimental Cell Research. 1999;**248**:97-109. DOI: 10.1006/excr.1999.4392.

[89] Fanning AS, Ma TY, Anderson JM. Isolation and functional characterization of the actin binding region in the tight junction protein ZO-1. FASEB Journal. 2002;**16**:1835-1837. DOI: 10.1096/fj.02-0121fje

[90] Molecular organization and function of invertebrate occluding junctions. Seminaries in Cell & Developmental Biology. 2014;**36**:186-193. DOI: 10.1016/j.semcdb.2014.09.009

[91] Suzuki H, Ito Y, Yamazaki Y, Mineta K, Uji M, Abe K, et al. The four-transmembrane protein IP39 of Euglena forms strands by a trimeric unit repeat. Nature Communications. 2013;**4**:1766. DOI: 10.1038/ncomms2731

[92] Cereijido M, Robbins ES, Dolan WJ, Rotunno CA, Sabatini DD. Polarized monolayers formed by epithelial cells on a permeable and translucent support. Journal of Cell Biology. 1978;**77**:853-880

[93] Boulpaep EL, Seely JF. Electrophysiology of proximal and distal tubules in the autoperfused dog kidney. American Journal of Physiology. 1971;**221**:1084-1096

[94] Lutz MD, Cardinal JE, Burg MB. Electrical resistance of renal proximal tubule perfused in vitro. American Journal of Physiology. 1973;**225**:729-734

[95] Helman SI, Grantham JJ, Burg MB. Effect of vasopressin on electrical resistance of renal cortical collecting tubules. American Journal of Physiology. 1971;**220**:1825-1832

[96] Rau WS, Frömter E. Electrical properties of the medullary collecting ducts of the golden hamster kidney. II. The transepithelial resistance. Pflügers Archiv. 1974;**351**:113-131

[97] Lavelle JP, Meyers SA, Ruiz WG, Buffington CT, Zeidel ML, Apodaca G. Urothelial pathophysiological changes in feline interstitial cystitis: A human model. American Journal of Physiology. Renal Physiology. 2000;**278**:F540-F553

[98] Lewis SA, Eaton DC, Clausen CH, Diamond JM. Nystatin as a probe for investigating the electrical properties of a tight epithelium. The Journal of General Physiology. 1977;**70**:427-440

[99] Furuse M, Furuse K, Sasaki H, Tsukita S. Conversion of zonulae occludentes from tight to leaky strand type by introducing claudin-2 into Madin-Darby canine kidney I cells. Journal of Cell Biology. 2001;**153**:263-272

[100] Amasheh S, Meiri N, Gitter AH, Schöneberg T, Mankertz J, Schulzke JD, et al. Claudin-2 expression induces cation-selective channels in tight junctions of epithelial cells. Journal of Cell Science. 2002;**115**:4969-4976

[101] Rosenthal R, Milatz S, Krug SM, Oelrich B, Schulzke JD, Amasheh S, et al., Claudin-2, a component of the tight junction, forms a paracellular water channel. Journal of Cell Science. 2010;**123**:1913-1921. DOI: 10.1242/jcs.060665

[102] Gonzalez-Mariscal L, Del CarmenNamorado M, Martin D, Sierra G, Reyes JL. The tight junction proteins claudin-7 and -8 display a different subcellular localization at Henle's loops and collecting ducts of rabbit kidney. Nephrology, Dialysis, Transplantation. 2006;**21**:2391-2398. DOI: 10.1093/ndt/gfl255

[103] Kiuchi-Saishin Y, Gotoh S, Furuse M, Takasuga A, Tano Y, Tsukita S. Differential expression patterns of claudins, tight junction membrane proteins, in mouse nephron segments. Journal of the American Society of Nephrology. 2002;**13**:875-886

[104] Reyes JL, Lamas M, Martin D, Namorado MD, Islas S, Luna J, et al. The renal segmental distribution of claudins changes with development. Kidney International. 2002;**62**:476-487. DOI: 10.1046/j.1523-1755.2002.00479.x

[105] Muto S, Hata M, Taniguchi J, Tsuruoka S, Moriwaki K, Saitou M, et al., Claudin-2-deficient mice are defective in the leaky and cation-selective paracellular permeability properties of renal proximal tubules. Proceedings of the National academy of Sciences of the United States of America. 2010;**107**:8011-8016. DOI: 10.1073/pnas.0912901107

[106] Colegio OR, Van Itallie C, Rahner C, Anderson JM. Claudin extracellular domains determine paracellular charge selectivity and resistance but not tight junction fibril architecture. American Journal of Physiology. Cell Physiology. 2003;**284**:C1346-C1354. DOI: 10.1152/ajpcell.00547.2002

[107] Colegio OR, Van Itallie CM, McCrea HJ, Rahner C, Anderson JM. Claudins create charge-selective channels in the paracellular pathway between epithelial cells. American Journal of Physiology. Cell Physiology. 2002;**283**:C142-C147. DOI: 10.1152/ajpcell.00038.2002

[108] Morita K, Furuse M, Fujimoto K, Tsukita S. Claudin multigene family encoding four-transmembrane domain protein components of tight junction strands. Proceedings of the National Academy of Sciences. 1999;**96**:511-516

[109] Gallardo JM, Hernández JM, Contreras RG, Flores-Maldonado C, González-Mariscal L, Cereijido M. Tight junctions are sensitive to peptides eliminated in the urine. Journal of Membrane Biology. 2002;**188**:33-42. DOI: 10.1007/s00232-001-0170-6

[110] Flores-Benitez D, Ruiz-Cabrera A, Flores-Maldonado C, Shoshani L, Cereijido M, Contreras RG. Control of tight junctional sealing: Role of epidermal growth factor. American Journal of Physiology. Renal Physiology. 2007;**292**:F828-F836. DOI: 10.1152/ajprenal.00369.2006

[111] Mullin JM, Laughlin KV, Ginanni N, Marano CW, Clarke HM, Peralta Soler A. Increased tight junction permeability can result from protein kinase C activation/translocation and act as a tumor promotional event in epithelial cancers. Annals of the New York Academy of Sciences. 2000;**915**:231-236

[112] Singh AB, Harris RC. Epidermal growth factor receptor activation differentially regulates claudin expression and enhances transepithelial resistance in Madin-Darby canine kidney cells. Journal of Biological Chemistry. 2004;**279**:3543-3552. DOI: 10.1074/jbc.M308682200

[113] Ikari A, Takiguchi A, Atomi K, Sugatani J. Epidermal growth factor increases clathrin-dependent endocytosis and degradation of claudin-2 protein in MDCK II cells. Journal of Cellular Physiology. 2011;**226**:2448-2456. DOI: 10.1002/jcp.22590

[114] Garcia-Hernandez V, Flores-Maldonado C, Rincon-Heredia R, Verdejo-Torres O, Bonilla-Delgado J, Meneses-Morales I, et al. EGF regulates claudin-2 and -4 expression through Src and STAT3 in MDCK cells. Journal of Cellular Physiology. 2015;**230**:105-115. DOI: 10.1002/jcp.24687

[115] Singh AB, Dhawan P. Claudins and cancer: Fall of the soldiers entrusted to protect the gate and keep the barrier intact. Seminars in Cell and Developmental Biology. 2015;**42**:58-65. DOI: 10.1016/j.semcdb.2015.05.001

[116] Ikari A, Atomi K, Takiguchi A, Yamazaki Y, Miwa M, Sugatani J. Epidermal growth factor increases claudin-4 expression mediated by Sp1 elevation in MDCK cells. Biochemical and Biophysical Research Communications. 2009;**384**:306-310. DOI: 10.1016/j.bbrc.2009.04.120

[117] Ikari A, Atomi K, Takiguchi A, Yamazaki Y, Hayashi H, Hirakawa J, et al. Enhancement of cell-cell contact by claudin-4 in renal epithelial Madin-Darby canine kidney cells. Journal of Cellular Biochemistry. 2012;**113**:499-507. DOI: 10.1002/jcb.23373

[118] Flores-Benitez D, Rincon-Heredia R, Razgado LF, Larre I, Cereijido M, Contreras RG. Control of tight junctional sealing: Roles of epidermal growth factor and prostaglandin E2. American Journal of Physiology. Cell Physiology. 2009;**297**:C611-C620. DOI: 10.1152/ajpcell.00622.2008

[119] Lu R, Johnson DL, Stewart L, Waite K, Elliott D, Wilson JM. Rab14 regulation of claudin-2 trafficking modulates epithelial permeability and lumen morphogenesis. Molecular Biology of the Cell. 2014;**25**:1744-1754. DOI: 10.1091/mbc.E13-12-0724

[120] Nighot PK, Hu CA, Ma TY. Autophagy enhances intestinal epithelial tight junction barrier function by targeting claudin-2 protein degradation. Journal of Biological Chemistry. 2015;**290**:7234-7246. DOI: 10.1074/jbc.M114.597492

[121] Yang Y, Li W, Sun Y, Han F, Hu CA, Wu Z. Amino acid deprivation disrupts barrier function and induces protective autophagy in intestinal porcine epithelial cells. Amino Acids. 2015;**47**:2177-2184. DOI: 10.1007/s00726-014-1844-6

[122] Zhou Y, Zheng B, Ye L, Zhang H, Zhu S, Zheng X, et al. Retinoic acid prevents disruption of blood-spinal cord barrier by inducing autophagic flux after spinal cord injury. Neurochemical Research. 2016;**41**:813-825. DOI: 10.1007/s11064-015-1756-1

[123] Withering W. An Account of the Foxglove, and Some of Its Medical Uses, 1st ed. Printed by N. Swinney for GGJ and J. Robinson, Paternoster-Row. London; 1785

[124] Schatzmann HJ. Herzglykoside als Hemmstoffe für den aktiven Kalium- und Natriumtransport durch die Erythrocytenmembran. Helvetica Physiologica et Pharmacologica Acta. 1953;**11**:346-354

[125] Bauer N, Müller-Ehmsen J, Krämer U, Hambarchian N, Zobel C, Schwinger RH, Neu H, Kirch U, Grünbaum EG, Schoner W. Ouabain-like compound changes rapidly on physical exercise in humans and dogs: Effects of beta-blockade and angiotensin-converting enzyme inhibition. Hypertension. 2005;**45**:1024-1028. DOI: 10.1161/01.HYP.0000165024.47728.f7

[126] Schneider R, Wray V, Nimtz M, Lehmann WD, Kirch U, Antolovic R, et al. Bovine adrenals contain, in addition to ouabain, a second inhibitor of the sodium pump. Journal of Biological Chemistry. 1998;**273**:784-792

[127] Schoner W, Bauer N, Müller-Ehmsen J, KrÄmer U, Hambarchian N, Schwinger R, et al. Ouabain as a mammalian hormone. Annals of the New York Academy of Sciences. 2003;**986**:678-684

[128] Schoner W, Scheiner-Bobis G. Endogenous cardiac glycosides: Hormones using the sodium pump as signal transducer. Seminars in Nephrology. 2005;**25**:343-351. DOI: 10.1016/j.semnephrol.2005.03.010

[129] Moreth K, Kuske R, Renner D, Schoner W. Blood pressure in essential hypertension correlates with the concentration of a circulating inhibitor of the sodium pump. Klinische Wochenschrift. 1986;**64**:239-244. DOI: 10.1007/BF01711656

[130] Fedorova OV, Shapiro JI, Bagrov AY. Endogenous cardiotonic steroids and salt-sensitive hypertension. Biochimica Et Biophysica Acta (BBA)—Molecular Basis of Disease. 2010;**1802**:1230-1236. DOI: 10.1016/j.bbadis.2010.03.011

[131] Nesher M, Shpolansky U, Viola N, Dvela M, Buzaglo N, Ben-Ami HC, et al. Ouabain attenuates cardiotoxicity induced by other cardiac steroids. British Journal of Pharmacology. 2010;**160**:346-354. DOI: 10.1111/j.1476-5381.2010.00701.x

[132] Koltsova SV, Trushina Y, Haloui M, Akimova OA, Tremblay J, Hamet P, et al. Ubiquitous [Na+]i/[K+]i-sensitive transcriptome in mammalian cells: Evidence for Ca(2+)i-independent excitation-transcription coupling. PLoS One. 2012;**7**:e38032. DOI: 10.1371/journal.pone.0038032

[133] Wang H, Haas M, Liang M, Cai T, Tian J, Li S, et al. Ouabain assembles signaling cascades through the caveolar Na+/K+-ATPase. Journal of Biological Chemistry. 2004;**279**:17250-17259. DOI: 10.1074/jbc.M313239200

[134] Aizman O, Uhlén P, Lal M, Brismar H, Aperia A. Ouabain, a steroid hormone that signals with slow calcium oscillations. Proceedings of the National Academy of Sciences. 2001;**98**:13420-13424. DOI: 10.1073/pnas.221315298

[135] Zhang S, Malmersjö S, Li J, Ando H, Aizman O, Uhlén P, et al. Distinct role of the N-terminal tail of the Na,K-ATPase catalytic subunit as a signal transducer. Journal of Biological Chemistry. 2006;**281**:21954-21962. DOI: 10.1074/jbc.M601578200

[136] Akimova OA, Hamet P, Orlov SN. [Na+]i/[K+]i-independent death of ouabain-treated renal epithelial cells is not mediated by Na+,K+-ATPase internalization and de novo gene expression. Pflügers Archiv. 2008;**455**:711-719. DOI: 10.1007/s00424-007-0283-6

[137] Contreras RG, Flores-Maldonado C, Lazaro A, Shoshani L, Flores-Benitez D, Larre I, et al. Ouabain binding to Na+,K+-ATPase relaxes cell attachment and sends a specific signal (NACos) to the nucleus. Journal of Membrane Biology. 2004;**198**:147-158. DOI: 10.1007/s00232-004-0670-2

[138] Liang M, Cai T, Tian J, Qu W, Xie ZJ. Functional characterization of Src-interacting Na/K-ATPase using RNA interference assay. Journal of Biological Chemistry. 2006;**281**:19709-19719. DOI: 10.1074/jbc.M512240200

[139] Larré I, Lazaro A, Contreras RG, Balda MS, Matter K, Flores-Maldonado C, et al. Ouabain modulates epithelial cell tight junction. Proceedings of the National academy of Sciences of the United States of America. 2010;**107**:11387-11392. DOI: 10.1073/pnas.1000500107

[140] Contreras RG, Shoshani L, Flores-Maldonado C, Lázaro A, Cereijido M. Relationship between Na(+),K(+)-ATPase and cell attachment. Journal of Cell Science. 1999;**112** (Pt 23):4223-4232

[141] Rincon-Heredia R, Flores-Benitez D, Flores-Maldonado C, Bonilla-Delgado J, García-Hernández V, Verdejo-Torres O, et al. Ouabain induces endocytosis and degradation of tight junction proteins through ERK1/2-dependent pathways. Experimental Cell Research. 2014;**320**:108-118. DOI: 10.1016/j.yexcr.2013.10.008

[142] Akimova OA, Tverskoi AM, Smolyaninova LV, Mongin AA, Lopina OD, La J, et al. Critical role of the α1-Na+, K+-ATPase subunit in insensitivity of rodent cells to cytotoxic action of ouabain. Apoptosis. 2015;**20**:1200-1210. DOI: 10.1007/s10495-015-1144-y

[143] Akimova OA, Tremblay J, Van Huysse JW, Hamet P, Orlov SN. Cardiotonic steroid-resistant alpha1-Na+,K+-ATPase rescues renal epithelial cells from the cytotoxic action of ouabain: Evidence for a Nai+,Ki+-independent mechanism. Apoptosis. 2010;**15**:55-62. DOI: 10.1007/s10495-009-0429-4

[144] Contreras RG, Flores-Beni TD, Flores-Maldonado C, Larre I, Shoshani L, Cereijido M. Na+,K+-ATPase and hormone ouabain:new roles for an old enzyme and an old inhibitor. Cellular and Molecular Biology (Noisy-le-Grand, France). 2006;**52**:31-40

[145] Hou J, Rajagopal M, Yu ASL. Claudins and the kidney. Annual Review of Physiology. 2013;**75**:479-501. DOI: 10.1146/annurev-physiol-030212-183705

[146] Arteaga ME, Hunziker W, Teo AS, Hillmer AM, Mutchinick OM. Familial hypomagnesemia with hypercalciuria and nephrocalcinosis: Variable phenotypic expression in three affected sisters from Mexican ancestry. Renal Failure. 2015;**37**:180-183. DOI: 10.3109/0886022X.2014.977141

[147] Sharma S, Place E, Lord K, Leroy BP, Falk MJ, Pradhan M. Claudin 19-based familial hypomagnesemia with hypercalciuria and nephrocalcinosis in a sibling pair. Clinical Nephrology. 2016;**85**:346-352. DOI: 10.5414/CN108783

[148] Hou J, Renigunta A, Konrad M, Gomes AS, Schneeberger EE, Paul DL, et al. Claudin-16 and claudin-19 interact and form a cation-selective tight junction complex. Journal of Clinical Investigation. 2008;**118**:619-628. DOI: 10.1172/JCI33970

[149] Hou J, Paul DL, Goodenough DA. Paracellin-1 and the modulation of ion selectivity of tight junctions. Journal of Cell Science. 2005;**118**:5109-5118. DOI: 10.1242/jcs.02631

[150] Gong Y, Renigunta V, Zhou Y, Sunq A, Wang J, Yang J, et al. Biochemical and biophysical analyses of tight junction permeability made of claudin-16 and claudin-19 dimerization. Molecular Biology of the Cell. 2015;**26**(24):4333-4346. DOI: 10.1091/mbc.E15-06-0422

[151] Kausalya PJ, Amasheh S, Günzel D, Wurps H, Müller D, Fromm M, et al. Disease-associated mutations affect intracellular traffic and paracellular Mg2+ transport function of Claudin-16. Journal of Clinical Investigation. 2006;**116**:878-891. DOI: 10.1172/JCI26323

[152] Müller D, Kausalya PJ, Meij IC, Hunziker W. Familial hypomagnesemia with hypercalciuria and nephrocalcinosis: Blocking endocytosis restores surface expression of a novel Claudin-16 mutant that lacks the entire C-terminal cytosolic tail. Human Molecular Genetics. 2006;**15**:1049-1058. DOI: 10.1093/hmg/ddl020

[153] Müller D, Kausalya PJ, Claverie-Martin F, Meij IC, Eggert P, Garcia-Nieto V, et al. A novel claudin 16 mutation associated with childhood hypercalciuria abolishes binding to ZO-1 and results in lysosomal mistargeting. The American Journal of Human Genetics. 2003;**73**:1293-1301. DOI: 10.1086/380418

[154] Ikari A, Matsumoto S, Harada H, Takagi K, Degawa M, Takahashi T, et al. Dysfunction of paracellin-1 by dephosphorylation in Dahl salt-sensitive hypertensive rats. Journal of Physiological Sciences. 2006;**56**:379-383. DOI: 10.2170/physiolsci.SC008906

[155] Cong X, Zhang Y, Li J, Mei M, Ding C, Xiang RL, et al. Claudin-4 is required for modulation of paracellular permeability by muscarinic acetylcholine receptor in epithelial cells. Journal of Cell Science. 2015;**128**:2271-2286. DOI: 10.1242/jcs.165878

[156] Mandai K, Nakanishi H, Satoh A, Obaishi H, Wada M, Nishioka H, et al. Afadin: A novel actin filament-binding protein with one PDZ domain localized at cadherin-based cell-to-cell adherens junction. Journal of Cell Biology. 1997;**139**:517-528

[157] Sakisaka T, Ikeda W, Ogita H, Fujita N, Takai Y. The roles of nectins in cell adhesions: Cooperation with other cell adhesion molecules and growth factor receptors. Current Opinion in Cell Biology. 2007;**19**:593-602. DOI: 10.1016/j.ceb.2007.09.007

[158] Kemler R, Ozawa M. Uvomorulin-catenin complex: Cytoplasmic anchorage of a Ca2+-dependent cell adhesion molecule. Bioessays. 1989;**11**:88-91. DOI: 10.1002/bies.950110403

[159] Hirano S, Takeichi M. Cadherins in brain morphogenesis and wiring. Physiological Reviews. 2012;**92**:597-634. DOI: 10.1152/physrev.00014.2011

[160] Nose A, Nagafuchi A, Takeichi M. Expressed recombinant cadherins mediate cell sorting in model systems. Cell. 1988;**54**:993-1001

[161] Takeichi M. Cadherin cell adhesion receptors as a morphogenetic regulator. Science. 1991;**251**:1451-1455

[162] Pokutta S, Weis WI. Structure and mechanism of cadherins and catenins in cell-cell contacts. Annual Review of Cell and Developmental Biology. 2007;**23**:237-261. DOI: 10.1146/annurev.cellbio.22.010305.104241

[163] Niessen CM, Gottardi CJ. Molecular components of the adherens junction. Biochimica et Biophysica Acta. 2008;**1778**:562-571. DOI: 10.1016/j.bbamem.2007.12.015

[164] Gumbiner BM. Regulation of cadherin-mediated adhesion in morphogenesis. Nature Reviews Molecular Cell Biology. 2005;**6**:622-634. DOI: 10.1038/nrm1699

[165] Abedin M, King N. The premetazoan ancestry of cadherins. Science. 2008;**319**:946-948. DOI: 10.1126/science.1151084

[166] Nanes BA, Chiasson-MacKenzie C, Lowery AM, Ishiyama N, Faundez V, Ikura M, et al. p120-catenin binding masks an endocytic signal conserved in classical cadherins. Journal of Cell Biology. 2012;**199**:365-380. DOI: 10.1083/jcb.201205029

[167] Anastasiadis PZ, Reynolds AB. The p120 catenin family: Complex roles in adhesion, signaling and cancer. Journal of Cell Science. 2000;**113**(Pt 8):1319-1334.

[168] Fujita Y, Krause G, Scheffner M, Zechner D, Leddy HEM, Behrens J, et al. Hakai, a c-Cbl-like protein, ubiquitinates and induces endocytosis of the E-cadherin complex. Nature Cell Biology. 2002;**4**:222-231. DOI: 10.1038/ncb758

[169] Petherick KJ, Williams AC, Lane JD, Ordóñez-Morán P, Huelsken J, Collard TJ, et al. Autolysosomal β-catenin degradation regulates Wnt-autophagy-p62 crosstalk. The EMBO Journal. 2013;**32**:1903-1916. DOI: 10.1038/emboj.2013.123

[170] Gugnoni M, Sancisi V, Gandolfi G, Manzotti G, Ragazzi M, Giordano D, et al. Cadherin-6 promotes EMT and cancer metastasis by restraining autophagy. Oncogene. 2017;**36**:667-677. DOI: 10.1038/onc.2016.237

[171] Pang M, Wang H, Rao P, Zhao Y, Xie J, Cao Q, et al. Autophagy links β-catenin and Smad signaling to promote epithelial-mesenchymal transition via upregulation of integrin linked kinase. International Journal of Biochemistry and Cell Biology. 2016;**76**:123-134. DOI: 10.1016/j.biocel.2016.05.010

[172] Goodenough DA, Goliger JA, Paul DL. Connexins, connexons, and intercellular communication. Annual Review of Biochemistry. 1996;**65**:475-502. DOI: 10.1146/annurev.bi.65.070196.002355

[173] Goodenough DA, Paul DL. Beyond the gap: Functions of unpaired connexon channels. Nature Reviews Molecular Cell Biology. 2003;**4**:285-294. DOI: 10.1038/nrm1072

[174] Hervé JC, Derangeon M. Gap-junction-mediated cell-to-cell communication. Cell and Tissue Research. 2013;**352**:21-31. DOI: 10.1007/s00441-012-1485-6

[175] Prochnow N, Dermietzel R. Connexons and cell adhesion: A romantic phase. Histochemistry and Cell Biology. 2008;**130**:71-77. DOI: 10.1007/s00418-008-0434-7

[176] Hervé JC, Bourmeyster N, Sarrouilhe D. Diversity in protein-protein interactions of connexins: Emerging roles. Biochimica et Biophysica Acta. 2004;**1662**:22-41. DOI: 10.1016/j.bbamem.2003.10.022

[177] Singh D, Solan JL, Taffet SM, Javier R, Lampe PD. Connexin 43 interacts with zona occludens-1 and -2 proteins in a cell cycle stage-specific manner. Journal of Biological Chemistry. 2005;**280**:30416-30421. DOI: 10.1074/jbc.M506799200

[178] Green KJ, Jones JC. Desmosomes and hemidesmosomes: Structure and function of molecular components. FASEB Journal. 1996;**10**:871-881

[179] Nekrasova O, Green KJ. Desmosome assembly and dynamics. Trends in Cell Biology. 2013;**23**:537-546. DOI: 10.1016/j.tcb.2013.06.004

[180] Getsios S, Huen AC, Green KJ. Working out the strength and flexibility of desmosomes. Nature Reviews Molecular Cell Biology. 2004;**5**:271-281. DOI: 10.1038/nrm1356

[181] Wallis S, Lloyd S, Wise I, Ireland G, Fleming TP, Garrod D. The alpha isoform of protein kinase C is involved in signaling the response of desmosomes to wounding in cultured epithelial cells. Molecular Biology of the Cell. 2000;**11**:1077-1092

[182] Garrod DR, Berika MY, Bardsley WF, Holmes D, Tabernero L. Hyper-adhesion in desmosomes: Its regulation in wound healing and possible relationship to cadherin crystal structure. Journal of Cell Science. 2005;**118**:5743-5754. DOI: 10.1242/jcs.02700

[183] Goodenough DA, Gilula NB. The splitting of hepatocyte gap junctions and zonulae occludentes with hypertonic disaccharides. Journal of Cell Biology. 1974;**61**:575-590

[184] Ghoshroy S, Goodenough DA, Sosinsky GE. Preparation, characterization, and structure of half gap junctional layers split with urea and EGTA. Journal of Membrane Biology. 1995;**146**:15-28

[185] Berthoud VM, Minogue PJ, Laing JG, Beyer EC. Pathways for degradation of connexins and gap junctions. Cardiovascular Research. 2004;**62**:256-267. DOI: 10.1016/j.cardiores.2003.12.021

[186] Bjorkman N. A study of the ultrastructure of the granulosa cells of the rat ovary. Acta Anatomica (Basel). 1962;**51**:125-147

[187] Carette D, Gilleron J, Denizot JP, Grant K, Pointis G, Segretain D. New cellular mechanisms of gap junction degradation and recycling. Biology of the Cell. 2015;**107**:218-231. DOI: 10.1111/boc.201400048

[188] Zou H, Zhuo L, Han T, Hu D, Yang X, Wang Y, et al. Autophagy and gap junctional intercellular communication inhibition are involved in cadmium-induced apoptosis in rat liver cells. Biochemical and Biophysical Research Communications. 2015;**459**:713-719. DOI: 10.1016/j.bbrc.2015.03.027

[189] Falk MM, Kells RM, Berthoud VM. Degradation of connexins and gap junctions. FEBS Letters. 2014;**588**(8):1221-1229. DOI: 10.1016/j.febslet.2014.01.031

[190] Bejarano E, Girao H, Yuste A, Patel B, Marques C, Spray DC, et al. Autophagy modulates dynamics of connexins at the plasma membrane in a ubiquitin-dependent manner. Molecular Biology of the Cell. 2012;**23**:2156-2169. DOI: 10.1091/mbc.E11-10-0844

[191] Garrod D, Tabernero L. Hyper-adhesion: A unique property of desmosomes. Cell Communication & Adhesion. 2014;**21**:249-256. DOI: 10.3109/15419061.2014.930133

[192] Allen TD, Potten CS. Desmosomal form, fate, and function in mammalian epidermis. Journal of Ultrastructure Research. 1975;**51**:94-105

[193] McHarg S, Hopkins G, Lim L, Garrod D. Down-regulation of desmosomes in cultured cells: The roles of PKC, microtubules and lysosomal/proteasomal degradation. PLoS One. 2014;**9**:e108570. DOI: 10.1371/journal.pone.0108570

[194] Calkins CC, Setzer SV, Jennings JM, Summers S, Tsunoda K, Amagai M, et al. Desmoglein endocytosis and desmosome disassembly are coordinated responses to pemphigus autoantibodies. Journal of Biological Chemistry. 2006;**281**:7623-7634. DOI: 10.1074/jbc.M512447200

[195] Barczyk M, Carracedo S, Gullberg D. Integrins. Cell and Tissue Research. 2010;**339**:269-280. DOI: 10.1007/s00441-009-0834-6

[196] Hynes RO. Integrins: Bidirectional, allosteric signaling machines. Cell. 2002;**110**:673-687

[197] Oommen S, Francois M, Kawasaki M, Murrell M, Kawasaki K, Porntaveetus T, et al. Cytoplasmic plaque formation in hemidesmosome development is dependent on SoxF transcription factor function. PLoS One. 2012;**7**:e43857. DOI: 10.1371/journal.pone.0043857

[198] Burridge K, Guilluy C. Focal adhesions, stress fibers and mechanical tension. Experimental Cell Research. 2016;**343**:14-20. DOI: 10.1016/j.yexcr.2015.10.029

[199] Wehrle-Haller B. Structure and function of focal adhesions. Current Opinion in Cell Biology. 2012;**24**:116-124. DOI: 10.1016/j.ceb.2011.11.001

[200] Paluch EK, Aspalter IM, Sixt M. Focal adhesion-independent cell migration. Annual Review of Cell and Developmental Biology. 2016;**32**:469-490. DOI: 10.1146/annurev-cellbio-111315-125341

[201] Tuloup-Minguez V, Hamaï A, Greffard A, Nicolas V, Codogno P, Botti J. Autophagy modulates cell migration and β1 integrin membrane recycling. Cell Cycle. 2013;**12**:3317-3328. DOI: 10.4161/cc.26298

[202] De Franceschi N, Hamidi H, Alanko J, Sahgal P, Ivaska J. Integrin traffic—the update. Journal of Cell Science. 2015;**128**:839-852. DOI: 10.1242/jcs.161653

[203] Vlahakis A, Jayanta D. The interconnections between autophagy and integrin-mediated cell adhesion. Journal of Molecular Biology. 2016:1-16. DOI: 10.1016/j.jmb.2016.11.027

[204] Kenific CM, Torsen W, Jayanta D. Autophagy in adhesion and migration. Journal of Cell Science. 2016;**129**:3685-3693. DOI: 10.1242/jcs.188490

Section 2

Techniques - Lysosomal Function

Chapter 6

Cellular Screening Methods for the Study of Nanoparticle-Induced Lysosomal Damage

Eleonore Fröhlich

Additional information is available at the end of the chapter

http://dx.doi.org/10.5772/intechopen.69306

Abstract

Nanoparticles (NPs) are included in many products of daily life and present in the environment. Due to the potential of NPs to improve quality and stability of consumer and health and medical products, it is expected that the exposure of humans to engineered NPs will rather increase than decrease in the future. Although NPs did not act acutely cytotoxic on these concentrations, they may cause adverse effects upon chronic exposure. Cytotoxicity testing in long-term cultures and analysis of organelle function could identify such effects. Cells take up NPs mainly via active mechanisms, and these routes deliver their payload predominantly to lysosomes. Acute exposure of cells to NPs can have adverse effects on lysosome morphology and function, but lysosomes are also potential targets for accumulation. The chapter explains the role of lysosomes and describes techniques for labeling and assessment of their function. Examples for co-localization studies and vital dye staining are shown. A variety of techniques are available to characterize effects of NPs on lysosomes, but care has to be taken in the choice of the proper technique because NPs may interfere with the detection.

Keywords: cathepsins, intralysosomal pH, staining, nanoparticles, toxicity, fluorescence microscopy

1. Introduction

Screening for acute cytotoxicity is an established procedure in the evaluation of chemicals, drugs, and medical devices and has also been used for the toxicological assessment of nanoparticles (NPs). Routine cytotoxicity testing detects changes in cell number, DNA, protein content, or metabolic activity of cells exposed to compounds applied in a broad concentration range. Exposure is usually for 4–72 h. In the case of conventional compounds, effects at high concentrations can indicate the toxicological potential of the compound at lower doses.

This prediction is not possible for NPs because particle agglomeration is more prominent at higher particle concentrations and agglomerates usually act less toxic than single NPs [1]. The exposure times of routine cytotoxicity testing are also less representative for NPs. Exposure to NPs occurs by contact with food, by the environment, and by consumer products in low doses but for prolonged time. The lack of good prediction of NP toxicity by acute testing protocols is also due to the fact that NPs in these products are usually poorly biodegradable and may accumulate in cells [2]. To address this problem, culture systems have been developed that allow the evaluation of cellular effects over prolonged time [3, 4]. Another option to identify toxicity upon prolonged contact is the study of organelles that are likely targets for damage by NPs. Active cellular uptake, endocytosis, represents the most common mechanism for cellular entry of NPs. Since the main active uptake routes deliver their payload to lysosomes, these organelles are the most likely targets for NP accumulation, potentially leading to lysosome dysfunction and cell damage.

2. Lysosomes

Lysosomes are cell organelles with an acidic lumen and a single outer membrane consisting of a phospholipid bilayer. They contain acid hydrolases, which enable the cell to process nutrients and destroy itself after death. Lysosomes are integrated in the mechanism of secretion and degradation of macromolecules and linked by vesicle transport to other intracellular structures, such as endosomes, the endoplasmic reticulum, and the Golgi apparatus (**Figure 1**).

Mammalian cells, with exception of erythrocytes, possess lysosomes, and some cell types also contain lysosome-related organelles, namely, melanosomes in melanocytes, lytic granules in cytotoxic T cells, delta granules in platelets, and lamellar bodies in alveolar epithelial cells. Lysosomes have spherical or tubular shape and measure <1 μm in non-phagocytic cells [5]. In phagocytes, which have the ability to ingest particles up to 10 μm, lysosomes can reach several micrometers in size [6]. Lysosomes are transported in the cytoplasm by passive and active mechanisms. While diffusion is size dependent and smaller lysosomes move faster than larger ones, active transport is independent from size [7].

Lysosomes digest macromolecules taken up by endocytosis (heterophagy), degrade intracellular macromolecules and organelles sequestered by autophagy, eliminate pathogens engulfed by phagocytosis, regulate metal ion levels, and sense nutrient availability. Lysosomal exocytosis is an unconventional secretion relevant for plasma membrane repair, immune response, and bone resorption [8]. Antigen processing by lysosomes is essential for the presentation of antigenic proteins to T cells [9, 10]. Lysosomes regulate the metabolic (anabolic or catabolic) state of the cells by sensing the nutritional state of the cell and conveying this information to the nucleus [11]. Transcription factor EB (TFEB), a master regulator of lysosomal biogenesis, co-localizes with mechanistic target of rapamycin complex 1 (mTORC1) on the lysosomal membrane. When nutrients are present, phosphorylation of TFEB by lysosomal surface-bound mTORC1 inhibits TFEB activity. The active mTORC1 promotes biosynthetic pathways and blocks autophagy. Catabolic pathways are switched on upon release (combined with inactivation)

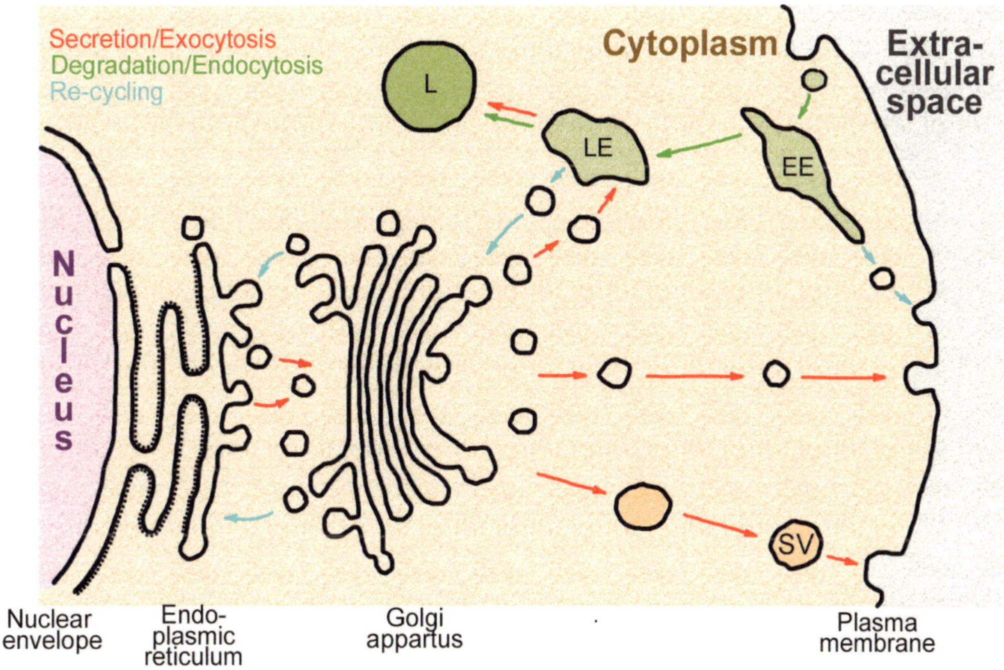

Figure 1. Integration of lysosomes in the cellular vesicle network. Ingested macromolecules or particles from the extracellular space reach the lysosomes (L) via early endosomes (EE) and late endosomes (LE) and can be degraded there. Macromolecules can be recycled through vesicle transfer to the Golgi apparatus and endoplasmic reticulum. Secretion products are synthesized at the rough endoplasmic reticulum and Golgi apparatus and are exported for constitutive secretion, like transport to the plasma membrane, or are included in secretory vesicles (SV) for regulated secretion into the extracellular space.

of bound mTORC1 from lysosomes in case of starvation or exercise, for instance. TFEB acts as a sensor of lysosomal state, when on the lysosomal surface, and as an effector of lysosomal function when translocated into the nucleus. Nuclear localization of TFEB is prevented when lysosome function is optimal. When TFEB has translocated into the nucleus, it upregulates the expression of genes encoding lysosomal proteins in order to improve lysosome function [12].

Normal lysosome function depends on an acid intralysosomal pH, and vacuolar/vesicular type H^+-ATPase (v-ATPase) is the most important proton pump for the regulation. Variation in pH between lysosomes of the same cell is considerable because lysosomes display functional and structural heterogeneity. By labeling lysosomes with a pH-insensitive and a pH-sensitive dye, Johnson et al. found that peripheral lysosomes had more alkaline pH values than perinuclear lysosomes [13]. The group also reported that the increase of the intralysosomal pH was linked to a change in the intracellular localization. The cellular volume of lysosomes is regulated in such a way that accumulation of undigested material induces increase in size and number of lysosomes to compensate reduced lysosome function.

Although lysosomes are involved in synthesis and recycling of macromolecules, their main role is usually seen in degradation. Degradation of extracellular material occurs after active uptake (**Figure 2**). With the exception of caveolin-dependent uptake, all routes deliver their cargo exclusively to lysosomes. Caveolin-mediated uptake can deliver macromolecules to the Golgi apparatus, endoplasmic reticulum, and lysosomes [14]. Endocytosis is relatively fast, and the maturation of endosomes to lysosomes takes approximately 40 min [5]. Lysosomes are also involved in the degradation of intracellular macromolecules and organelles, which occurs as microautophagy, chaperone-mediated autophagy, and macroautophagy (usually referred to as autophagy). Autophagy includes the following steps: vesicle nucleation (phagophore formation), vesicle expansion (autophagosome formation), maturation (fusion with multivesicular bodies (MVBs) or lysosomes), and degradation [15, 16]. Degradation can only occur when lysosomes are active. The increased presence of autophagosomes may be due to excessive induction of autophagy or to blockade of autosome degradation (autophagy flux).

Macropinocytosis, clathrin-mediated uptake, caveolin-mediated, and clathrin- and caveolin-independent uptake ingest NPs although with different size preferences (for more detail, see, for instance, Ref. [17]). Particles larger than 500 nm are taken up by phagocytosis (**Figure 3**).

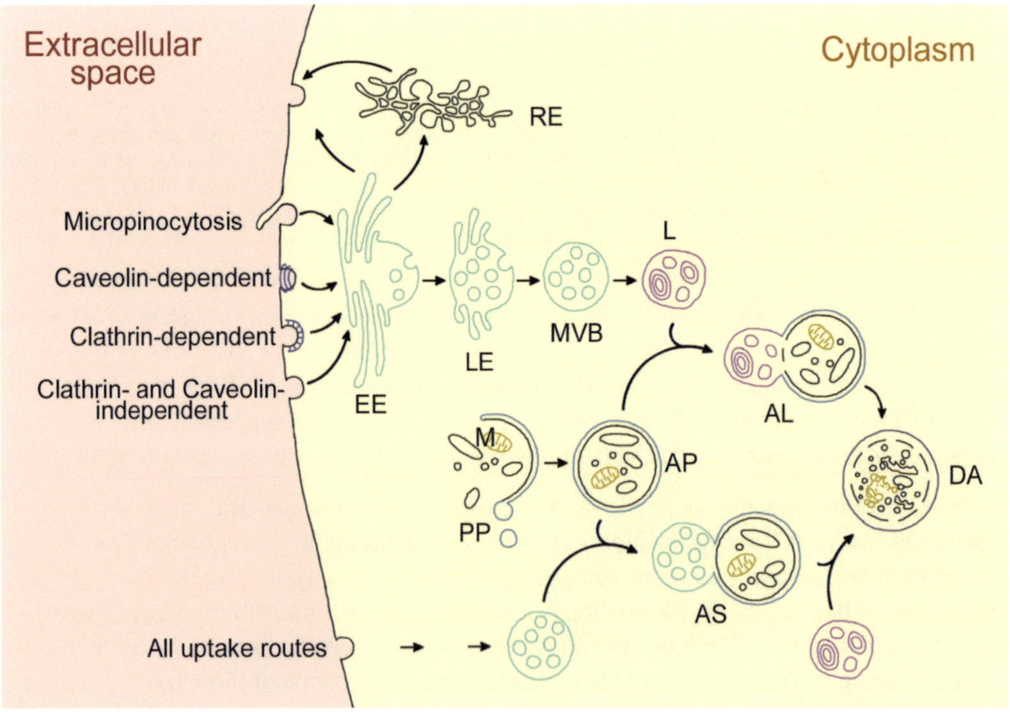

Figure 2. Interaction of lysosomes with uptake routes and autophagy. Payload of the active uptake routes micropinocytosis, caveolin-dependent, clathrin-dependent, and clathrin- and caveolin-independent routes is delivered to early endosomes (EEs), late endosomes (LEs), multivesicular bodies (MVBs), and lysosomes (Ls). EEs can recycle to the plasma membrane as recycling endosomes (REs). Macroautophagy is started by the formation of the phagophore (PP) and forms the autophagosome (AP). PPs can contain parts of cytoplasm and organelles, for instance, mitochondria (M). APs may fuse with MVBs to build amphisomes (ASs) or with Ls to form autolysosomes (ALs). ALs mature to degrading autolysosomes (DAs), which can also arise by the fusion of ASs with Ls.

http://dx.doi.org/10.5772/intechopen.69306

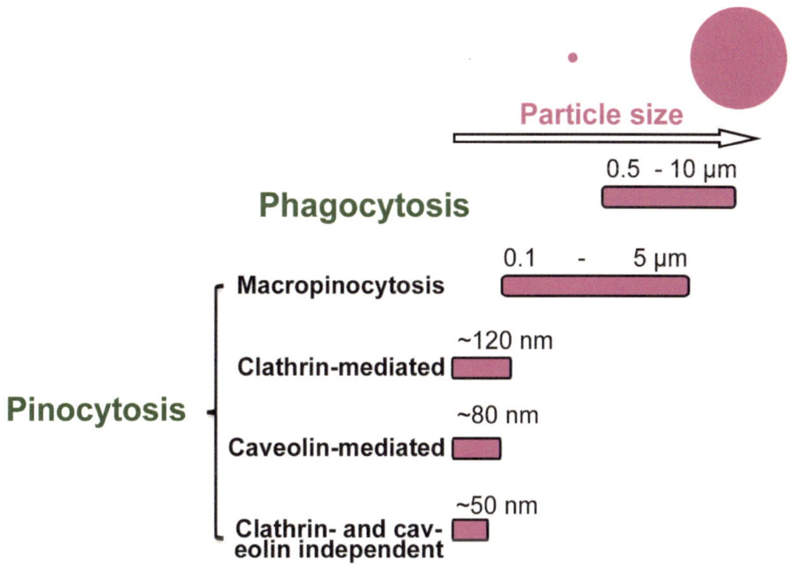

Figure 3. Overview of size preferences of the different active uptake routes for particles.

Silica (SiO_2), gold, iron oxide, polystyrene particles, and quantum dots have been detected in lysosomes in the absence of obvious morphological damage [18–28]. They can reach the lysosomes by different uptake routes, and it appears that particle size, and cell-specific expression of the different uptake routes determines the preferential uptake mechanism [29]. Cellular excretion of NPs is estimated to be low, and therefore, accumulation of NPs in lysosomes and chronic impairment of lysosome function may occur [30].

3. Symptoms and causes of lysosome dysfunction

Garnett and Kallinteri [31] suggested that accumulation of NPs might interfere with lysosomal function and cause similar symptoms as lysosomal storage diseases. The deficiency of specific lysosomal enzymes, mainly lysosomal sulfatases arylsulfatase A, B, and G, causes lysosomal storage diseases [32] with neurological, pulmonary, and cardiac impairment [33]. Lysosomal dysfunction, however, can also be caused by accumulation of pharmaceutical compounds. Several drugs, such as nonsteroidal anti-inflammatory drugs, statins, antidepressants, beta-blockers, tyrosine kinase inhibitors, anti-histamines, and so on, are sequestered in lysosomes and may cause drug-induced lysosomal damage, termed phospholipidosis. Phagocytic cells, alveolar and peritoneal macrophages, are more sensitive than non-phagocytic cells, and damage causes the histopathological image of "foamy macrophages" [34]. The intracellular accumulation of membranous material shows analogies to inherited lysosomal storage disorder [35] and suggests that external factors may also cause analogous symptoms. Molecules, which cause lysosomal damage, are typically organic amines, such as chloroquine, amiodarone, perhexiline, aminoglycosides, and chlorphentermine. These compounds accumulate in endosomes and can cause swelling and disruption of lysosomes with subsequent cell death [36].

Impaired lysosomal function may also lead to abnormal autophagy. The role of autophagy in diseases is not completely clear. Reduced autophagy promotes cell transformation and development of cancer, whereas in later stages, increased autophagy allows the cancer cells to survive regardless of deprivation of nutrients. Disruption of autophagy has also a promoting role in neurodegenerative diseases [37]. Drugs act on autophagy mainly by alteration of lysosomal enzyme activity and of intralysosomal pH [38]. The epoxysuccinyl compound E64d and Pepstatin A inhibit cysteine and aspartic proteases, whereas chloroquine, hydroxychloroquine, Lys05, NH_4Cl, matrine, momensine, and lucanthone increase intralysosomal pH, and bafilomycin A1, azithromycin, and concanamycin A inhibit v-ATPase.

Inactive non-digested material as part of normal biological aging can accumulate in the form of the autofluorescent pigment lipofuscin. This accumulation of lipids, carbohydrates, and aldehyde-crosslinked proteins increases cellular susceptibility to oxidative stress, alters intralysosomal pH and lysosomal membrane permeability, and impairs lysosomal function [39].

Accumulation in lysosomes has a prominent influence on the elimination of drugs from the body [40]. Based on the assumption that drugs that are trapped in lysosomes (lysosomotropic drugs) and non-lysosomotropic drugs have a plasma concentration of 1mg/L and no physical binding to blood or tissues occurs, the distribution volume of non-lysosomotropic drugs is 42L for the average 70kg person. To determine the distribution volume of lysosomotropic drugs, the lysosome volume has to be added. It was calculated as 0.5 L based on the assumption that the body contains 15 trillion cells with a volume of 3.4×10^{-9} cm^3 (average hepatocyte volume) and 1% of which represent lysosomes. To be factored into the total body volume, drug concentration must be the same as in plasma. The apparent lysosome volume, therefore, is 500L, which means ~10 times greater than the distribution volume of a non-lysosomotropic drug. This estimation appears not unrealistic since accumulation of chloroquine in rat tissues at typical therapeutic concentrations was 800 higher than values in plasma, and cells were able to accumulate propranolol 1000-fold compared to extracellular concentrations. Based on these calculations, a 10 times longer half-life for lysosomotropic compared to non-lysosomotropic drugs is expected. A long persistence in tissues has also been reported for NPs in several studies (e.g., in Refs. [41, 42]).

4. Lysosomal damage by nanoparticles

In contrast to nuclear membrane and plasma membrane, lysosomes possess only a single phospholipid bilayer for resistance against the around 60 different intralysosomal hydrolases. More than 50% of the lysosomal membrane proteins of late endosomes and lysosomes consist of lysosome-associated membrane proteins (LAMPs). LAMPs are the most densely glycosylated proteins (>60% of total mass) and form the inner lining of the lysosomal membrane. It is postulated that the composition with high glycosylation and low content of cholesterol represents a better protection against the action of lysosomal hydrolases than the composition of conventional membranes [9].

Despite the good protection against hydrolases, lysosomal membranes can be acutely damaged by NPs [43, 44]. Depending on the extent of the damage, cell death via different mechanisms ensues [45]. Partial permeabilization, in general, results in reactive oxygen species (ROS) generation and apoptotic cell death, whereas massive permeabilization induces cytosolic acidification and necrosis. Iron oxide NPs induced cytotoxicity through generation of ROS by lysosomes. Toxic ions released by partial degradation of particles in lysosomes were the main toxic mechanism of zinc oxide (ZnO) and copper oxide (CuO) NPs [46]. Quantum dots in lysosomes caused swelling of lysosomes linked to morphological alterations [47], and cationic cerium oxide (CeO_2) NPs and polystyrene particles induced disruption of lysosomes [48]. As toxic mechanism of cationic NPs, buffering of H^+ with increased lysosomal pH is assumed [49]. Multiwalled carbon nanotubes (CNTs) increased lysosomal permeability by direct action on lysosomal membranes [50]. The increase in autophagosomes is a common finding in the cellular action of NPs, such as fullerenes, gold NPs, iron oxide NPs, rare-earth oxide NPs, quantum dots, CNTs, titanium dioxide (TiO_2) NPs, and SiO_2 NPs [51–61]. In addition to decreased phagosome degradation, impaired lysosomal trafficking by disruption of the actin cytoskeleton may be a reason for the increased presence of autophagosomes [37]. Since all NPs, despite different compositions and surface functionalities, increased cellular autophagosome content, it is suspected that increase in autophagy is inherently linked to the small size of these particles.

Subtle changes in lysosome physiology without obvious morphological alterations have also been reported. Exposure to 15–200 nm TiO_2 and 10 nm ZnO NPs induced changes in pH and enzyme activities in epithelial cells [43, 62, 63], while polystyrene and TiO_2 NPs increased intralysosomal pH of macrophages [62, 64]. In addition to increasing intralysosomal pH, polystyrene particles interfered with lysosomal enzyme activity [65]. By increasing the intralysosomal pH, silver NPs decreased the pH-dependent uptake of fluorescent indicator dyes [66]. Increased expression of CatB protein was observed in SiO_2 NP-treated cells [51]. Based on the finding that accumulation of undigested material leads to upregulation of lysosome size and number [5, 67, 68], the increase of lysosomal activity may represent a compensation mechanism for impaired lysosome function caused by SiO_2 NPs.

5. Marker for interaction with lysosomes and its morphology, integrity, and function

A panel of methods is available to assess lysosome morphology and function. Area measurements can be performed by image analysis of (fluorescence-labeled) lysosomes. Morphological changes can be identified by transmission electron microscopy (TEM). Lysosomal function can be determined by changes in the expression of lysosome-related genes or proteins, while fluorescent substrates or fluorescent dyes indicate changes in enzyme activities or pH. Not all assays, however, are suitable for the assessment of NPs because colorimetric, fluorescent, and luminescent assays are prone to interference with NPs [69]. Interference can cause false-positive and false-negative results. In cytotoxicity testing with assays, where enzymatic activity is determined by absorbance of a colored product, colored NPs can mask cell loss because the absorbance caused by NPs leads to an overestimation of viable cells. Examples for overestimation of cell damage are membrane permeability assays with fluorescence-based assays. The increase of the signal by NPs with inherent fluorescence indicates more cell damage than actually present. High concentrations of NPs, on the other hand, may quench the fluorescent signal, and it is often appropriate to compare reader data with microscopical observation (**Figure 4**). In general, the parallel assessment by several techniques (plate reader and microscopy) and the switch to another detection method (fluorescence instead of absorbance) help to avoid false conclusions due to interference.

Microscopic techniques are frequently used for the assessment of NPs because they allow the correlation of intracellular localization and cellular effects. Examination by TEM cannot only reveal organelle damage but can also be used for co-localization studies because NPs contained in food, consumer products, and cosmetics can generally be visualized by TEM. Fluorescence microscopy using life stains and immunocytochemistry serves for area measurements. In combination with fluorescent particles, co-localization studies and intracellular tracking can be performed. Cells transduced/transfected with fluorescent protein-LAMP-1 constructs can also be used for these studies. The commercially available technology Organelle Lights™ uses a targeted fluorescent protein with viral delivery (http://web.mit.edu/rkarimi/www/Special/Other/Protocol/Organelle%20Lights_%20Intracellular%20Targeted%20Fluorescent%20Proteins.pdf) for transduction of mammalian cells. By using LAMP-1 transfected cells, it was found that small 20 nm carboxyl-functionalized polystyrene particles were preferentially located in the perinuclear region, whereas 200 nm particles were detected to a greater extent in the cellular periphery (**Figure 5**). Given the fact that perinuclear lysosomes have a more acidic pH than peripheral lysosomes [13], the uptake in different types of lysosomes may cause different cellular effects. For co-localization with lysosomes also immunocytochemical detection with antibodies against LAMP-1, LAMP-2, and LAMP-3 can be used.

Lysosome markers, such as gold-coupled albumin and fluorescence-labeled dextran, use active uptake for the labeling [70]. Since NPs may interfere with active uptake routes, these markers are less suitable for NP studies. Information on lysosome function can be obtained by detection of enzymatic activity or pH-dependent dyes.

Activity measurement of acid phosphatase, β-glucuronidase, and β-hexosaminidase, which have been released from lysosomes, can be used as marker for lysosome function but needs isolation of the organelles. NPs that were located outside the lysosomes can get access to the assay compounds during the isolation procedure and cause artificial effects. Such interference occurred when cathepsin B (CatB) activity of cells exposed to polystyrene particles was detected in homogenates [65]. In situ assays, where cells are not homogenized, can avoid this problem because only NPs located inside lysosomes get access to the substrate. For quantification of enzyme activity, in situ substrates for sulfatases and cathepsins are available. The substrate SulfGreen is metabolized by all lysosomal sulfatases (http://www.markergene.com/product_sheets/pis1377.pdf) and fluorescent substrates, for instance, CV-$(RR)_2$ for CatB and MR-$(FR)_2$ for cathepsin L indicate protease activity [65, 71].

Changes in intralysosomal pH value can be studied using pH-dependent dyes (acridine orange and neutral red retention). Acridine orange has been used for many years to visualize organelles with acidic pH. The dye stains lysosomes in green and red fluorescence at low extracellular concentrations (2.6 μM). When the concentration of the dye in the staining solution is higher (26–37 μM), stacks can be formed and lysosomes show red fluorescence. Unfortunately, the red stacks bleach very fast and only the monomeric (green) form

http://dx.doi.org/10.5772/intechopen.69306

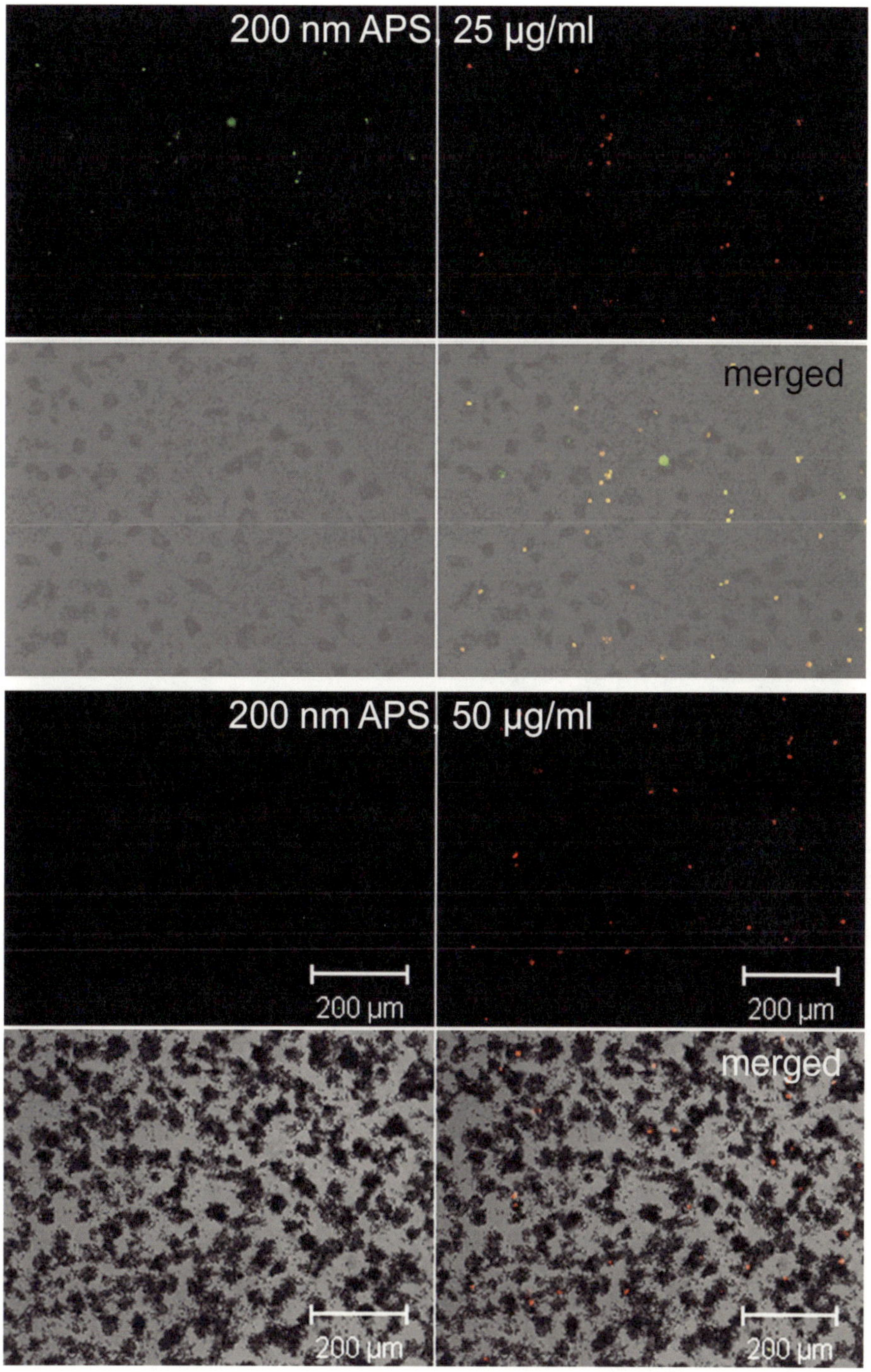

Figure 4. Detection of apoptosis and necrosis caused by 200 nm amidine-functionalized polystyrene (APS) particles in EAhy926 endothelial cells Staining in the different channels (green, upper left; red, upper right; bright field, lower left) and overlay (merged, lower right) is shown. Despite particle sedimentation (dark precipitates), the staining with YoPro-1 (green channel, upper left) for apoptosis and propidium iodide (red channel, upper right) for necrosis is visible. Lower concentrations (25 μg/mL) of the particles caused apoptosis and necrosis, whereas the higher concentration (50 μg/mL) induced only necrotic cell death.

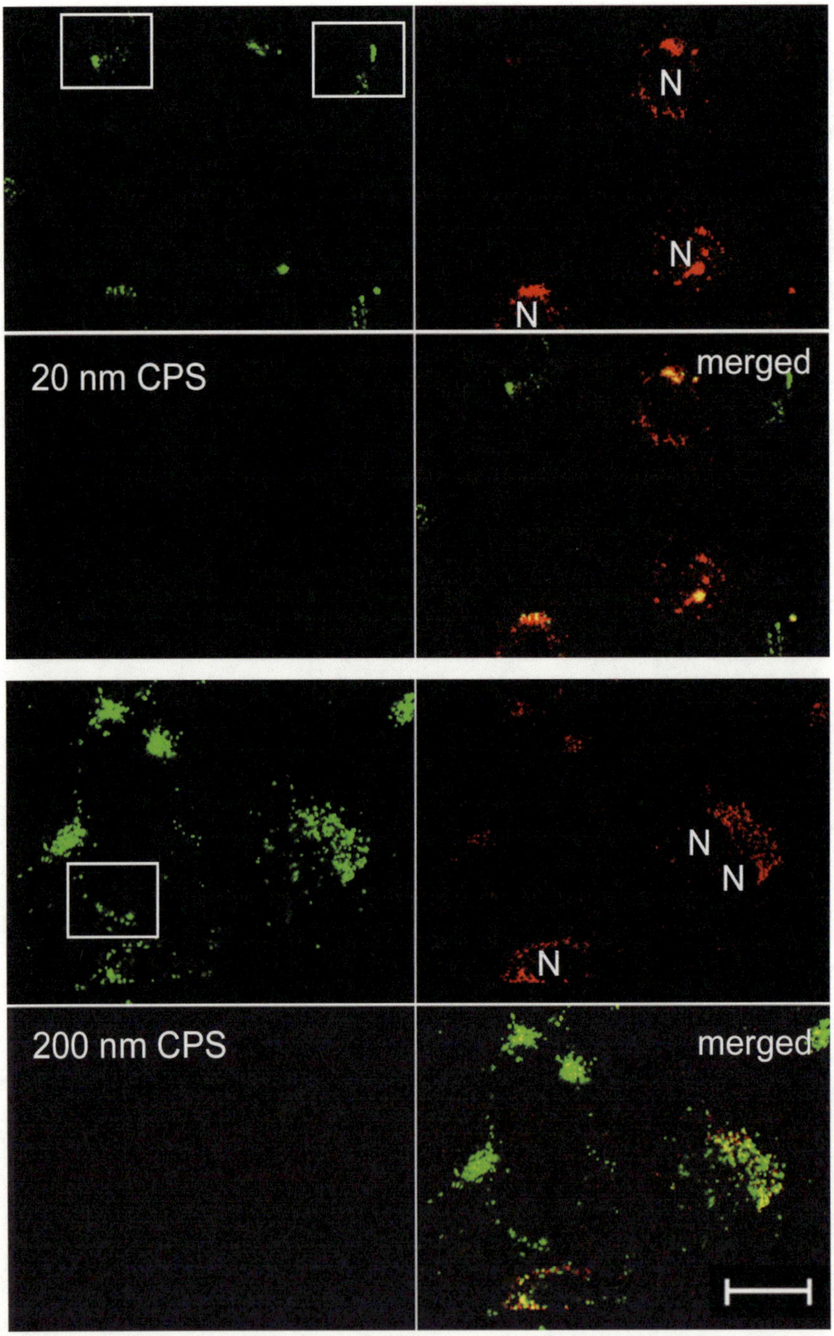

Figure 5. Confocal fluorescence images of 20 nm and 200 nm green-yellow fluorescently labeled carboxyl-functionalized (green) polystyrene NPs in RFP-LAMP-1 transfected EAhy926 endothelial cells (red). Staining in the different channels (green, upper left; red, upper right) and overlay (merged, lower right) is shown. Lysosomes are seen throughout the cells with exclusion of the nucleus (N). Uptake in non-transfected cells (indicated with squares in the green channel) was not obviously different from transfected cells. Small polystyrene particles (a) co-localized to a greater extent with perinuclear lysosomes than larger (b) particles. Abbreviation: RFP, red fluorescent protein. Scale bar, 20 μm.

remains [72]. In addition to that, acridine orange is phototoxic and induces burst of dye-loaded vesicles. Quinacrine also accumulates in lysosomes with acid pH, but the loading is accompanied with morphological alterations (swelling), which makes the staining less suitable for physiological studies. The probe LysoTracker™ Red DND-99, a lipophilic amine with logP 2.10 and pK_a 7.5, is another marker for accumulation in lysosomes [73]. Although it acts not markedly phototoxic, its usefulness as lysosome marker is limited by bleaching. Neutral red can be used as indicator for functional lysosomes [74] but is less sensitive than fluorescent dyes and more often used as viability screening test. Membrane permeant dyes, such as LysoTracker™ and LysoSensor™ probes, label lysosomes in living cells [75]. They are more selective than the classical neutral red and acridine orange dyes. The compound commercialized as Lyso-ID® is a cationic amphiphilic tracer that accumulates in acidic organelles. It can be used as indication for lysosome size and number and is used in the drug screening for lysosomal damage [76] (**Figure 6a, b**). Increased staining indicates swelling of lysosomes and increase in lysosome number. This increase is seen as adaptation to insufficient intracellular degradation capacity.

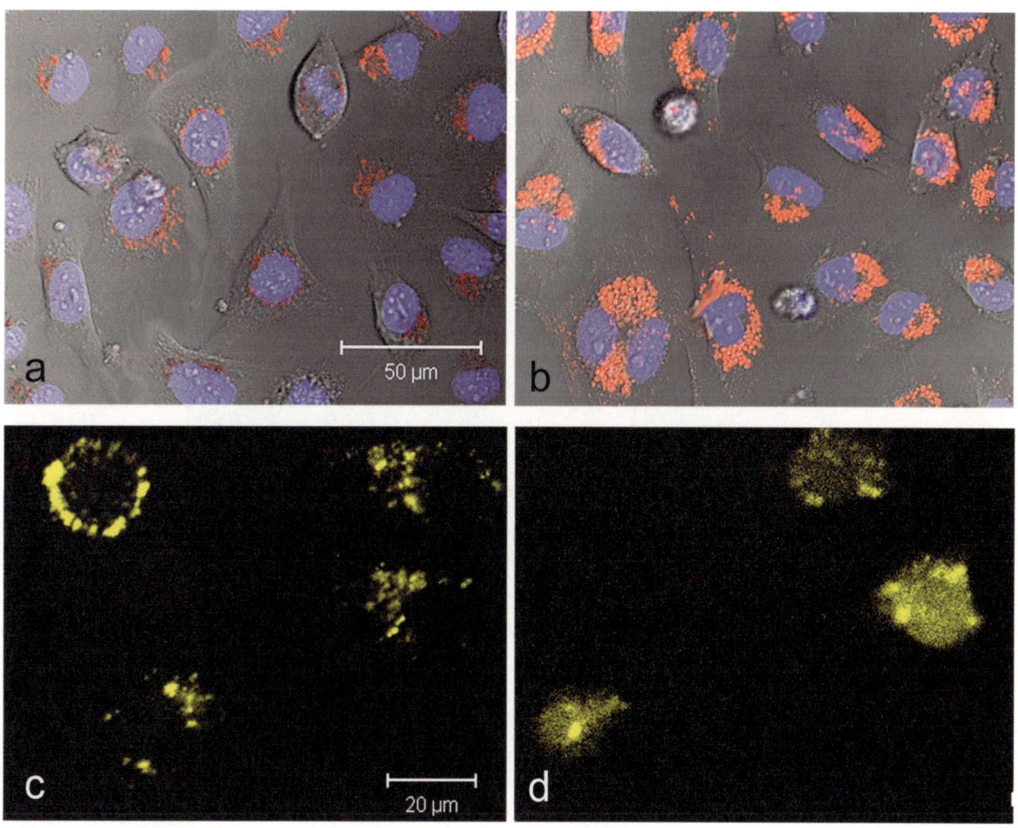

Figure 6. Lyso-ID staining in the cytoplasm (red channel) of untreated EAhy926 cells (a) is low but increased when cells are exposed to 25 µM chloroquine (b). Nuclei are stained with Hoechst 33342 (blue channel). Lysosomes also increased in size upon chloroquine treatment. Massive lysosome damage changed the staining pattern with Lucifer yellow in the cytoplasm from punctuate in untreated cells (c) to diffuse in chloroquine-treated cells (d).

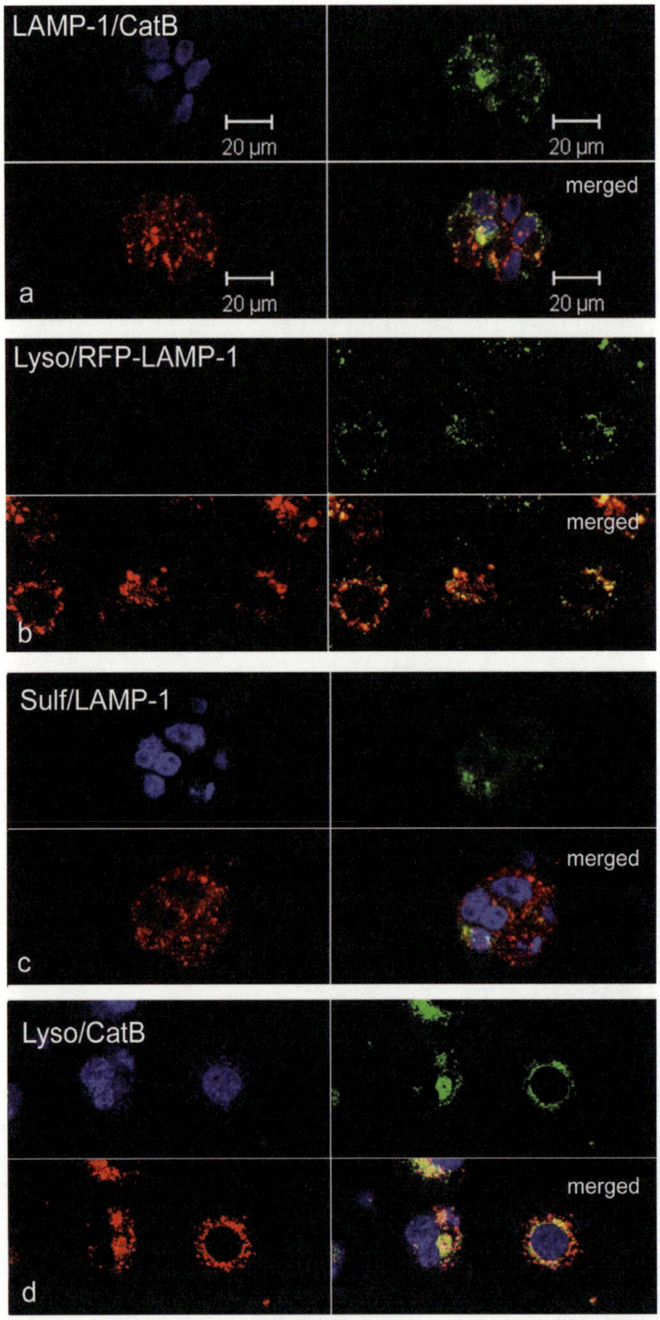

Figure 7. Co-staining with markers for enzyme activity and pH demonstrates interlysosomal heterogeneity of untreated EAhy926 endothelial cells. Lysosomes irrespective of their functional activity are identified by immunocytochemical staining with anti-lysosome-associated membrane protein 1 (LAMP-1) antibody or transduction with RFP-LAMP-1. Staining in the different channels (blue, upper left; green, upper right; red, lower left) and overlay (merged, lower right) is shown. The nuclear stain with Hoechst 33342 is seen in the blue channel. (a) LAMP-1-immunoreactivity (green channel)/CatB staining (red channel), (b) LysoSensor (Lyso) staining (green channel)/RFP-LAMP-1 (red channel), No nuclear staining has been performed. (c) Sulf staining (green channel)/LAMP-1-immunoreactivity (red channel), and (d) Lyso (green channel)/CatB (red channel) staining. Abbreviations: CatB, cathepsin B; LAMP-1, lysosome-associated membrane protein 1; Lyso, LysoSensor dye; RFP, red fluorescent protein; and Sulf, lysosomal sulfatases.

Finally, alterations of the staining pattern with Lucifer yellow from punctuate to diffuse staining can identify lysosome damage. The dye is accumulated and retained in healthy lysosomes but leaks out when the integrity of lysosome membranes is lost (**Figure 6c, d**).

Co-staining with combinations of different lysosomal proteins, pH, and enzyme activity can reveal the heterogeneity of lysosomes and may identify changes in amount and quality of lysosomes. In EAhy926 endothelial cells, LAMP-1-immunoreactive (ir)/CatB-negative lysosomes were seen more frequently than CatB-positive/LAMP-1-not ir structures (**Figure 7a**) and RFP-LAMP-1 positive/Lyso-negative more frequent than Lyso-positive/RFP-LAMP-1-negative structures (**Figure 7b**). Similarly, LAMP-1-ir/Sulf-negative structures were more often seen than Sulf-positive/LAMP-1-not ir structures (**Figure 7c**). This findings can be explained by the fact that RFP-LAMP-1 and LAMP-1 antibodies label all late endosomes and lysosomes, but low pH and prominent enzyme activity are only present in a particular subgroup. LysoSensor and CatB activity stain did also not completely co-localize. CatB-positive vesicles were located more at the cell periphery, while LysoSensor-positive structures were preferentially located in the perinuclear region (**Figure 7d**). This corresponds to the finding that lysosomes with low pH are preferentially located in the perinuclear region [13]. Activity of cathepsin L is also mainly seen in the perinuclear region, but CatB is active over a broader range of pH [77] and can be detected also in peripheral lysosomes.

Lysosomal activity is important for the execution of autophagy, and cellular increase of autophagosomes may indicate impaired lysosomal function. Microtubule-associated protein 1A/1B light chain 3 (LC3) is a cytosolic protein, which, during formation of autophagosomes, is conjugated to phosphatidylethanolamine. The conjugate is first recruited to autophagosomal membranes and, after fusion of autophagosomes with lysosomes, degraded by lysosomal proteases [78]. Immunoblotting or immunocytochemical detection of LC3 has become the most common screening marker for autophagy. Increase of LC3 immunoreactivity indicates an increased content of autophagosomes. Despite the strong link to lysosomes, LC3 and LAMP-1 are not co-localized (**Figure 8**).

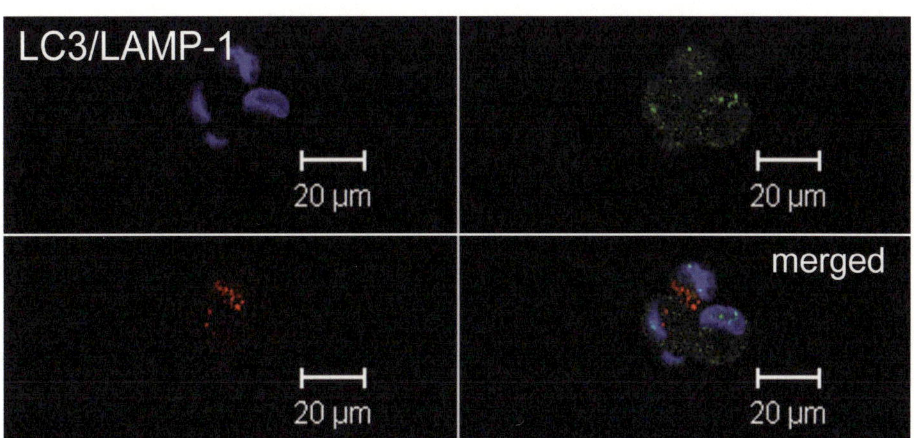

Figure 8. Co-localization of LC3-ir and LAMP-1-ir structures in EAhy926 endothelial cells. Staining in the different channels (blue, nuclei, upper left; green, LC3- ir, upper right; red, LAMP-1-ir, lower left) and overlay (merged, lower right) is shown. Abbreviations: LC3, microtubule-associated protein 1A/1B light chain 3 and LAMP-1, lysosome-associated membrane protein 1.

6. Conclusions

Based on the existing literature, lysosomes may be acutely damaged by high concentrations of NPs. Due to the accumulation in the endosomal-lysosomal system, it is supposed that NPs can damage lysosomes upon prolonged exposure. TEM can identify morphological changes, and a panel of vital stains allows the determination of intralysosomal pH and activity of lysosomal enzymes. While TEM analysis is time-consuming and lacks physiological information, immunocytochemical staining combined with cellular life stains is a good option to study lysosome function. In combination with fluorescent (labeled) NPs, uptake and localization in different parts of the endosomal-lysosomal system can be shown. Given the heterogeneity of lysosomes, particle localization in peripheral or perinuclear lysosomes may cause different cellular effects.

Acknowledgements

Support of the studies by the Austrian Science Fund grant P 22576-B18 is gratefully acknowledged.

Author details

Eleonore Fröhlich

Address all correspondence to: eleonore.froehlich@medunigraz.at

Medical University of Graz, Center for Medical Research, Graz, Austria

References

[1] Gogotsi Y. Nanomaterials handbook. In: Gogotsi Y, editor. Advanced Materials and Technologies. Boca Raton: CRC Press; 2006

[2] Simon-Deckers A, Gouget B, Mayne-L'hermite M, Herlin-Boime N, Reynaud C, Carriere M. In vitro investigation of oxide nanoparticle and carbon nanotube toxicity and intracellular accumulation in A549 human pneumocytes. Toxicology. 2008;**253**(1-3):137-146

[3] Mrakovcic M, Absenger M, Riedl R, Smole C, Roblegg E, Fröhlich L, Fröhlich E. Assessment of long-term effects of nanoparticles in a microcarrier cell culture system. PLoS One. 2013;**8**(2):e56791

[4] Mrakovcic M, Meindl C, Roblegg E, Fröhlich E. Reaction of monocytes to polystyrene and silica nanoparticles in short-term and long-term exposures. Toxicology Research. 2014;**3**:86-97

[5] Appelqvist H, Waster P, Kagedal K, Ollinger K. The lysosome: From waste bag to potential therapeutic target. Journal of Molecular Cell Biology. 2013;**5**(4):214-226

[6] Green TR, Fisher J, Stone M, Wroblewski BM, Ingham E. Polyethylene particles of a 'critical size' are necessary for the induction of cytokines by macrophages in vitro. Biomaterials. 1998;**19**(24):2297-2302

[7] Bandyopadhyay D, Cyphersmith A, Zapata JA, Kim YJ, Payne CK. Lysosome transport as a function of lysosome diameter. PLoS One. 2014;**9**(1):e86847

[8] Ballabio A. The awesome lysosome. EMBO Molecular Medicine. 2016;**8**(2):73-76

[9] Eskelinen EL, Tanaka Y, Saftig P. At the acidic edge: Emerging functions for lysosomal membrane proteins. Trends in Cell Biology. 2003;**13**(3):137-145

[10] Settembre C, Fraldi A, Medina DL, Ballabio A. Signals from the lysosome: A control centre for cellular clearance and energy metabolism. Nature Reviews, Molecular Cell Biology. 2013;**14**(5):283-296

[11] Settembre C, Zoncu R, Medina DL, Vetrini F, Erdin S, Erdin S, Huynh T, Ferron M, Karsenty G, Vellard MC, Facchinetti V, Sabatini DM, Ballabio A. A lysosome-to-nucleus signalling mechanism senses and regulates the lysosome via mTOR and TFEB. EMBO Journal. 2012;**31**(5):1095-1108

[12] Roczniak-Ferguson A, Petit CS, Froehlich F, Qian S, Ky J, Angarola B, Walther TC, Ferguson SM. The transcription factor TFEB links mTORC1 signaling to transcriptional control of lysosome homeostasis. Science Signaling. 2012;**5**(228):ra42

[13] Johnson DE, Ostrowski P, Jaumouille V, Grinstein S. The position of lysosomes within the cell determines their luminal pH. Journal of Cell Biology. 2016;**212**(6):677-692

[14] Le PU, Nabi IR. Distinct caveolae-mediated endocytic pathways target the Golgi apparatus and the endoplasmic reticulum. Journal of Cell Science. 2003;**116**(Pt 6):1059-1071

[15] Eskelinen EL, Saftig P. Autophagy: A lysosomal degradation pathway with a central role in health and disease. Biochimical and Biophysical Acta. 2009;**1793**(4):664-673

[16] Lee JA. Neuronal autophagy: A housekeeper or a fighter in neuronal cell survival? Experimental Neurobiology. 2012;**21**(1):1-8

[17] Sahay G, Alakhova DY, Kabanov AV. Endocytosis of nanomedicines. Journal of Controlled Release. 2010;**145**(3):182-195

[18] Al-Rawi M, Diabate S, Weiss C. Uptake and intracellular localization of submicron and nano-sized SiO(2) particles in HeLa cells. Archives of Toxicology. 2011;**85**(7):813-826

[19] He Q, Zhang Z, Gao Y, Shi J, Li Y. Intracellular localization and cytotoxicity of spherical mesoporous silica nano- and microparticles. Small. 2009;**5**(23):2722-2729

[20] Silver J, Ou W. Photoactivation of quantum dot fluorescence following endocytosis. Nano Letters. 2005;**5**(7):1445-1449

[21] Lévy R, Shaheen U, Cesbron Y, Sée V. Gold nanoparticles delivery in mammalian live cells: A critical review. Nano Reviews. 2010;**1**

[22] Chithrani BD, Ghazani AA, Chan WC. Determining the size and shape dependence of gold nanoparticle uptake into mammalian cells. Nano Letters. 2006;**6**(4):662-668

[23] Faklaris O, Joshi V, Irinopoulou T, Tauc P, Sennour M, Girard H, Gesset C, Arnault JC, Thorel A, Boudou JP, Curmi PA, Treussart F. Photoluminescent diamond nanoparticles for cell labeling: Study of the uptake mechanism in mammalian cells. ACS Nano. 2009;**3**(12):3955-3962

[24] Goya GF, Marcos-Campos I, Fernandez-Pacheco R, Saez B, Godino J, Asin L, Lambea J, Tabuenca P, Mayordomo JI, Larrad L, Ibarra MR, Tres A. Dendritic cell uptake of iron-based magnetic nanoparticles. Cell Biology International. 2008;**32**(8):1001-1005

[25] Jaiswal JK, Mattoussi H, Mauro JM, Simon SM. Long-term multiple color imaging of live cells using quantum dot bioconjugates. Nature Biotechnology. 2003;**21**(1):47-51

[26] Nativo P, Prior IA, Brust M. Uptake and intracellular fate of surface-modified gold nanoparticles. ACS Nano. 2008;**2**(8):1639-1644

[27] Rejman J, Oberle V, Zuhorn IS, Hoekstra D. Size-dependent internalization of particles via the pathways of clathrin- and caveolae-mediated endocytosis. Biochemical Journal. 2004;**377**(Pt 1):159-169

[28] Stearns RC, Paulauskis JD, Godleski JJ. Endocytosis of ultrafine particles by A549 cells. American Journal of Respiratory Cell and Molecular Biology. 2001;**24**(2):108-115

[29] Kuhn DA, Vanhecke D, Michen B, Blank F, Gehr P, Petri-Fink A, Rothen-Rutishauser B. Different endocytotic uptake mechanisms for nanoparticles in epithelial cells and macrophages. Beilstein Journal of Nanotechnology. 2014;**5**:1625-1636

[30] Fröhlich E. Cellular elimination of nanoparticles. Environmental Toxicology and Pharmacology. 2016;**46**:90-94

[31] Garnett MC, Kallinteri P. Nanomedicines and nanotoxicology: Some physiological principles. Occupational Medicine. 2006;**56**(5):307-311

[32] Frese MA, Schulz S, Dierks T. Arylsulfatase G, a novel lysosomal sulfatase. Journal of Biological Chemistry. 2008;**283**(17):11388-11395

[33] Greiner-Tollersrud O, Berg T. Lysosomal storage disorders. In: Saftig P, editor. Lysosomes. New York: Springer Science, Landes Bioscience; 2005. pp. 60-73

[34] Van Bambeke F, Saffran J, Mingeot-Leclercq MP, Tulkens PM. Mixed-lipid storage disorder induced in macrophages and fibroblasts by oritavancin (LY333328), a new glycopeptide antibiotic with exceptional cellular accumulation. Antimicrobial Agents and Chemotherapy. 2005;**49**(5):1695-1700

[35] Anderson N, Borlak J. Drug-induced phospholipidosis. FEBS Letters. 2006;**580**(23):5533-5540

[36] van Dyke R. part III The lysosome in its cytoplasmic environment. Acidification of endosomes and lysosomes. Lysosomal membrane bodies. In: Lloyd J, Mason R, editors. Subcellular Biochemistry. Vol. 27 Biology of the Lysosome. London: New York Plenum Press; 1996

[37] Stern ST, Adiseshaiah PP, Crist RM. Autophagy and lysosomal dysfunction as emerging mechanisms of nanomaterial toxicity. Particle and Fibre Toxicology. 2012;**9**:20

[38] Vakifahmetoglu-Norberg H, Xia HG, Yuan J. Pharmacologic agents targeting autophagy. Journal of Clinical Investigation. 2015;**125**(1):5-13

[39] Orenstein S, Cuervo A, Changes in lysosomes and their autophagic function in aging: The comparative biology of lysosomal function. In: Wolf N, editor. Comparative Biology of Aging. Dordrecht: Springer Science+Business Media; 2010. pp. 201-226

[40] Hamid N, Krise JP, The Mechanisms and therapeutic consequences of amine-containing drug sequestration in lysosomes. In: Maxfield F, Willard J, Lu S, editors. Lysosomes: Biology, Diseases, and Therapeutics. Hoboken: John Wiley & Sons; 2016

[41] Geraets L, Oomen AG, Krystek P, Jacobsen NR, Wallin H, Laurentie M, Verharen HW, Brandon EF, de Jong WH. Tissue distribution and elimination after oral and intravenous administration of different titanium dioxide nanoparticles in rats. Particle and Fibre Toxicology. 2014;**11**:30

[42] Sadauskas E, Danscher G, Stoltenberg M, Vogel U, Larsen A, Wallin H. Protracted elimination of gold nanoparticles from mouse liver. Nanomedicine: Nanotechnology, Biology, and Medicine. 2009;**5**(2):162-169

[43] Cho WS, Duffin R, Howie SE, Scotton CJ, Wallace WA, Macnee W, Bradley M, Megson IL, Donaldson K. Progressive severe lung injury by zinc oxide nanoparticles; the role of Zn^{2+} dissolution inside lysosomes. Particle and Fibre Toxicology. 2011;**8**:27

[44] Jin CY, Zhu BS, Wang XF, Lu QH. Cytotoxicity of titanium dioxide nanoparticles in mouse fibroblast cells. Chemical Research in Toxicology. 2008;**21**(9):1871-1877

[45] Koehler A, Marx U, Broeg K, Bahns S, Bressling J. Effects of nanoparticles in *Mytilus edulis* gills and hepatopancreas - a new threat to marine life? Marine Environmental Research. 2008;**66**(1):12-14

[46] Vandebriel R, De Jong W. A review of mammalian toxicity of ZnO nanoparticles. Nanotechnology, Science and Applications. 2012;**5**:61-71

[47] Funnell WR, Maysinger D. Three-dimensional reconstruction of cell nuclei, internalized quantum dots and sites of lipid peroxidation. Journal of Nanobiotechnology. 2006; **4**(10):10

[48] Bexiga MG, Varela JA, Wang F, Fenaroli F, Salvati A, Lynch I, Simpson JC, Dawson KA. Cationic nanoparticles induce caspase 3-, 7- and 9-mediated cytotoxicity in a human astrocytoma cell line. Nanotoxicology. 2011;**5**(4):557-567

[49] Nel AE, Madler L, Velegol D, Xia T, Hoek EM, Somasundaran P, Klaessig F, Castranova V, Thompson M. Understanding biophysicochemical interactions at the nano-bio interface. Nature Materials. 2009;**8**(7):543-557

[50] Sohaebuddin SK, Thevenot PT, Baker D, Eaton JW, Tang L. Nanomaterial cytotoxicity is composition, size, and cell type dependent. Particle and Fibre Toxicology. 2010;**7**:22

[51] Halamoda Kenzaoui B, Chapuis Bernasconi C, Guney-Ayra S, Juillerat-Jeanneret L. Induction of oxidative stress, lysosome activation and autophagy by nanoparticles in human brain-derived endothelial cells. Biochemical Journal. 2012;**441**(3):813-821

[52] Khan MI, Mohammad A, Patil G, Naqvi SA, Chauhan LK, Ahmad I. Induction of ROS, mitochondrial damage and autophagy in lung epithelial cancer cells by iron oxide nanoparticles. Biomaterials. 2012;**33**(5):1477-1488

[53] Liu HL, Zhang YL, Yang N, Zhang YX, Liu XQ, Li CG, Zhao Y, Wang YG, Zhang GG, Yang P, Guo F, Sun Y, Jiang CY. A functionalized single-walled carbon nanotube-induced autophagic cell death in human lung cells through Akt-TSC2-mTOR signaling. Cell Death & Disease. 2011;**2**:e159

[54] Ma X, Wu Y, Jin S, Tian Y, Zhang X, Zhao Y, Yu L, Liang XJ. Gold nanoparticles induce autophagosome accumulation through size-dependent nanoparticle uptake and lysosome impairment. ACS Nano. 2011;**5**(11):8629-8639

[55] Stern ST, Zolnik BS, McLeland CB, Clogston J, Zheng J, McNeil SE. Induction of autophagy in porcine kidney cells by quantum dots: A common cellular response to nanomaterials? Toxicological Sciences. 2008;**106**(1):140-152

[56] Yamawaki H, Iwai N. Cytotoxicity of water-soluble fullerene in vascular endothelial cells. American Journal of Physiology Cell Physiology. 2006;**290**(6):C1495-C1502

[57] Zhang Q, Yang W, Man N, Zheng F, Shen Y, Sun K, Li Y, Wen LP. Autophagy-mediated chemosensitization in cancer cells by fullerene C60 nanocrystal. Autophagy. 2009;**5**(8):1107-1117

[58] Johnson-Lyles DN, Peifley K, Lockett S, Neun BW, Hansen M, Clogston J, Stern ST, McNeil SE. Fullerenol cytotoxicity in kidney cells is associated with cytoskeleton disruption, autophagic vacuole accumulation, and mitochondrial dysfunction. Toxicology and Applied Pharmacology. 2010;**248**(3):249-258

[59] Li JJ, Hartono D, Ong CN, Bay BH, Yung LY. Autophagy and oxidative stress associated with gold nanoparticles. Biomaterials. 2010;**31**(23):5996-6003

[60] Li C, Liu H, Sun Y, Wang H, Guo F, Rao S, Deng J, Zhang Y, Miao Y, Guo C, Meng J, Chen X, Li L, Li D, Xu H, Li B, Jiang C. PAMAM nanoparticles promote acute lung injury by inducing autophagic cell death through the Akt-TSC2-mTOR signaling pathway. Journal of Molecular Cell Biology. 2009;**1**(1):37-45

[61] Yu L, Lu Y, Man N, Yu SH, Wen LP. Rare earth oxide nanocrystals induce autophagy in HeLa cells. Small. 2009;**5**(24):2784-2787

[62] Hamilton RF, Wu N, Porter D, Buford M, Wolfarth M, Holian A. Particle length-dependent titanium dioxide nanomaterials toxicity and bioactivity. Particle and Fibre Toxicology. 2009;**6**:35

[63] Hussain S, Thomassen LC, Ferecatu I, Borot MC, Andreau K, Martens JA, Fleury J, Baeza-Squiban A, Marano F, Boland S. Carbon black and titanium dioxide nanoparticles elicit distinct apoptotic pathways in bronchial epithelial cells. Particle and Fibre Toxicology. 2010;**7**:10

[64] Lunov O, Syrovets T, Loos C, Beil J, Delacher M, Tron K, Nienhaus GU, Musyanovych A, Mailander V, Landfester K, Simmet T. Differential uptake of functionalized polystyrene nanoparticles by human macrophages and a monocytic cell line. ACS Nano. 2011;**5**(3):1657-1669

[65] Fröhlich E, Meindl C, Roblegg E, Ebner B, Absenger M, Pieber TR. Action of polystyrene nanoparticles of different sizes on lysosomal function and integrity. Particle and Fibre Toxicology. **2012**;9:26

[66] Miyayama T, Matsuoka M. Involvement of lysosomal dysfunction in silver nanoparticle-induced cellular damage in A549 human lung alveolar epithelial cells. Journal of Occupational Medicine and Toxicology. 2016;**11**:1

[67] Xu H, Ren D. Lysosomal physiology. Annual Review of Physiology. 2015;**77**:57-80

[68] Bendiske J, Bahr BA. Lysosomal activation is a compensatory response against protein accumulation and associated synaptopathogenesis--an approach for slowing Alzheimer disease? Journal of Neuropathology and Experimental Neurology. 2003;**62**(5):451-463

[69] Fröhlich E, Meindl C, Pieber T. Important issues in the cytotoxicity screening of nano-sized materials. EURO-NanoTox Letters. 2010;**1**:1-6

[70] Vogel SM, Minshall RD, Pilipovic M, Tiruppathi C, Malik AB. Albumin uptake and transcytosis in endothelial cells in vivo induced by albumin-binding protein. American Journal of Physiology. Lung Cellular and Molecular Physiology. 2001;**281**(6):L1512-L1522

[71] Pacheco FJ, Servin J, Dang D, Kim J, Molinaro C, Daniels T, Brown-Bryan TA, Imoto-Egami M, Casiano CA. Involvement of lysosomal cathepsins in the cleavage of DNA topoisomerase I during necrotic cell death. Arthritis Rheum. 2005;**52**(7):2133-2145

[72] Pierzynska-Mach A, Janowski PA, Dobrucki JW. Evaluation of acridine orange, LysoTracker Red, and quinacrine as fluorescent probes for long-term tracking of acidic vesicles. Cytometry. Part A. 2014;**85**(8):729-737

[73] Duvvuri M, Krise JP. A novel assay reveals that weakly basic model compounds concentrate in lysosomes to an extent greater than pH-partitioning theory would predict. Molecular Pharmacology. 2005;**2**(6):440-448

[74] Chazotte B. Labeling lysosomes in live cells with neutral red. Cold Spring Harbor Protocols. 2011;**2011**(2):pdb.prot5570

[75] Chazotte B. Labeling lysosomes in live cells with LysoTracker. Cold Spring Harbor Protocols. 2011;**2011**(2):pdb.prot5571

[76] Coleman J, Xiang Y, Pande P, Shen D, Gatica D, Patton WF. A live-cell fluorescence microplate assay suitable for monitoring vacuolation arising from drug or toxic agent treatment. Journal of Biomolecular Screening. 2010;**15**(4):398-405

[77] Turk V, Stoka V, Vasiljeva O, Renko M, Sun T, Turk B, Turk D. Cysteine cathepsins: From structure, function and regulation to new frontiers. Biochimical and Biophysical Acta. 2012;**1824**(1):68-88

[78] Tanida I, Ueno T, Kominami E. LC3 and autophagy. Methods in Molecular Biology. 2008;**445**:77-88

Chapter 7

Diagnostic Application of Lysosomal Exoglycosidases

Sylwia Chojnowska, Alina Kępka,
Slawomir Dariusz Szajda,
Napoleon Waszkiewicz and Krzysztof Zwierz

Additional information is available at the end of the chapter

http://dx.doi.org/10.5772/intechopen.69307

Abstract

Lysosomal exoglycosidases gradually degrade oligosaccharide chains of glycoconjugates (glycoproteins, glycolipids, glycosaminoglycans) in cell lysosomes. Defect in the activity of suitable lysosomal exoglycosidase stops degradation of oligosaccharide chains on sugar molecules not released by deficient exoglycosidase, and nondegraded oligosaccharide chains are stored in enlarged lysosomes. Enlarged lysosomes damage remaining cell structures and disturb the function of involved tissues, causing storage diseases. An increase in the activity of exoglycosidases in tissues and body fluids is observed in the reconstruction of damaged tissues. Exoglycosidase activity is an inexpensive and sensitive marker in diagnostics and monitoring of many diseases.

Keywords: lysosomal exoglycosidases, fucosidase (FUC), β-D-galactosidase (GAL), β-D-glucuronidase (GLU), N-acetyl-β-hexosaminidase (HEX), α- and β-mannosidases (MAN)

1. Introduction: lysosomes

Inside lysosomes, more than 50 hydrolytic enzymes (glycosidases, proteases, lipases, nucleases, phosphatases, sulfatases, etc.) that are able to degrade all types of cell macromolecules are located. Lysosomal enzymes are active at acidic (pH ~5.0) water environment. High intralysosomal [H^+] (about 100x higher than in cytoplasm) is maintained by vacuolar H^+, V-type ATPase, located in the lysosomal membrane, which uses the energy of ATP hydrolysis to pump protons into lysosomes [1–3]. Lysosomal unique highly acidic environment creates some sort of protection for cytoplasmic components against noncontrolled autodigestion, additionally reinforced by integral proteins of the lysosomal membrane that are highly glycosylated to protect both lysosomal membrane and cytosolic elements against autodigestion [2, 4, 5]. Furthermore, some of the membrane glycoproteins function as specific receptors for molecules destined to degradation in lysosomes [3].

Designated for autodigestion, extracellular high-molecular substances reach lysosomes by endocytosis, pinocytosis, and phagocytosis [6]. Intracellular high-molecular substances are digested by autophagy [7]; autophagy eliminates waste or damaged parts of the living cells. There are many types of autophagy: macroautophagy, microautophagy, chaperone-dependent autophagy, and specific autophagy. Autophagy may also be classified according to the digested material, e.g., mitophagy (digestion of mitochondria) or nucleophagy (digestion of nuclear debris) [8–10]. The best described is macroautophagy, where the cellular region destined for digestion is surrounded by the phospholipid membrane creating autophagosome. Then autophagosome merges with lysosome, where acid hydrolases degrade autosome contents into simple organic compounds, ready for utilization by the cell [11]. Additionally, autophagy provides the cells with energy [12]. Autophagy may be induced by hypoxia-caused stress, hunger, radiation, inflammation, and so on [8]. In the case of pathological autophagy, cells exposed to intracellular toxins suffer from defective metabolism and die. Some of the researchers suspect that deficient autophagy may initiate many diseases such as diabetes or Alzheimer's disease [8], or even cancerogenesis. On the other hand, excessive autophagy may facilitate the survival of neoplastic cells during harmful conditions (e.g. chemotherapy). Therefore, autophagy in neoplasia may have dual biological sense [13].

2. Lysosomal enzymes

In autophagy, lysosomal acid proteases and glycosidases play a main role. Proteases cleave peptide bonds in the middle (endopeptidases) or outside (exopeptidases) of polypeptide chains. Main group of lysosomal proteolytic enzymes constitute cathepsins [14, 15], having aspartate (cathepsin D and E), cysteine (cathepsins B, C, H, K, and L), or serine (cathepsins A and G) in the active site [16, 17]. Proteases (PROT) (**Figure 1**) facilitate the action of three groups of glycosidases that gradually degrade tissue glycoconjugates (glycoproteins, glycolipids, and glycosaminoglycans): **aminohydrolases** that hydrolyze the N-glycosidic linkage between amino group of polypeptide chain asparagine and N-acetylglucosamine of oligosaccharide chain of glycoprotein, **endoglycosidases** that hydrolyze O-glycosidic bonds inside of oligosaccharide chains, and **exoglycosidases** that hydrolyze O-glycosidic bonds releasing sugars from nonreducing and reducing terminals of oligosaccharide chains [18].

Aminohydrolases as well as endo- and exoglycosidases create a sequence of reactions where the product of the previous enzyme is the substrate for the subsequent enzyme (**Figure 1**), and oligosaccharide is digested from reducing and nonreducing ends. When neuraminidase (NEU) releases N-acetylneuraminic acid (NANA) from the nonreducing ends of oligosaccharide chains, PROT degrade protein cores of glycoproteins, releasing reducing ends of oligosaccharides with attached asparagines. Oligosaccharides deprived of NANA are substrates for appropriate exoglycosidases depending on oligosaccharide composition.

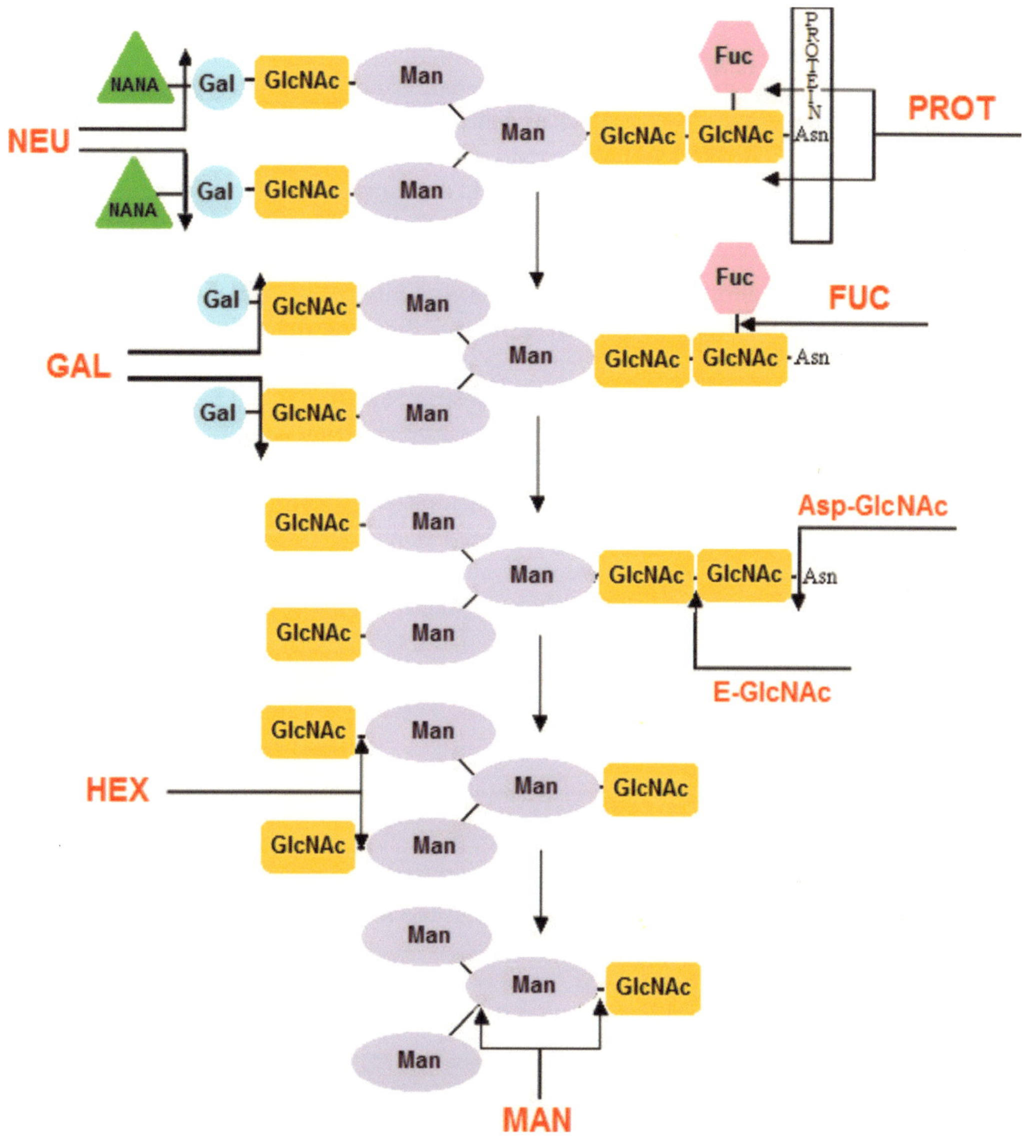

Figure 1. A sequential degradation of glycoproteins in lysosomes based on Ref. [19]. **Asn**—asparagine, **Fuc—**fucose, **Gal—**galactose, **GlcNAc**—N-acetylglucosamine, **Man**—mannose, **NANA**—N-acetylneuraminic acid, **Asp-GlucNAc**—aspartylglucosaminidase, **E-GlcNAc**—endo N-acetylglucosaminidase, **FUC**—α-fucosidase, **GAL**—β-galactosidase, **HEX**—N-acetyl-β-hexosaminidase, **MAN**—α- and β-mannosidases, **NEU**—neuraminidase (sialidase), **PROT**—proteases.

Oligosaccharides with β-D- galactose on non-reducing ends are substrates for β- galactosidase and oligosaccharides with α-L-fucose near reducing ends are substrates for α-L-fucosidase (**Figure 1**). Then, oligosaccharide chains are degraded by aspartylglucosaminidase (Asp-GlcNAc) that hydrolyses N-glycosidic bond between N-acetylglucosamine of the reducing end of oligosaccharide and asparagine remained from polypeptide as well as endo-N-acetylglucosaminidase (E-GlcNAc) releasing N-acetyloglucosamine from the reducing end of oligosaccharide chains. N-acetylhexosaminidase (NAG, N-acetyl-β-hexosaminidase (HEX)) releases N-acetyloglucosamine and N-acetylgalactosamine from a nonreducing end of the remaining part of oligosaccharide chains. Oligosaccharides containing mannose are substrates of α- and β – mannosidases (**Figure 1**). Lack or deficiency of a particular exoglycosidase blocks catabolism of oligosaccharide chains on a nondetached sugar residue [19]. Disorders in the activity of lysosomal enzymes are closely related to autophagy and reflect intensity of development and course of many diseases, for example, infections, inflammations, cancers, heart diseases, Crohn's disease, myopathy, liver diseases, and neurodegenerations. Autophagy is induced in cells by numerous factors: bacterial or viral infections, oxidative stress, and lack of nutrients. Some of the literature data also indicate the protective effect of autophagy [8, 20–23]. Increase in the activity of exoglycosidases in tissues [24–27] and body fluids [28–32] is observed in autophagy combined with the reconstruction of damaged tissues. In addition, determination of exoglycosidase activity is inexpensive and sensitive [33]. In joint diseases (osteoarthritis, rheumatoid arthritis, and Lyme arthritis), progressive destruction of joint cartilages occurs. Destruction of cartilage is a multifactorial process caused by concerted action of lysosomal hydrolases (**Figure 2**). Proteases digest polypeptide chains of glycoconjugates exposing glycopeptides. Endoglycosidases (hyaluronidases, chondroitinases, keratanases, etc.) break down glycosidic bonds inside glycoconjugates and release oligosaccharide chains from the protein core. Lysosomal exoglycosidases, HEX, GAL, β-D-glucuronidase (GLU), and so on, release monosaccharides from the nonreducing terminals of oligosaccharide chains of glycoproteins, glycolipids, and glycosaminoglycans of synovial tissue, articular cartilage, and synovial fluid (**Figure 2**) [34].

2.1. Lysosomal exoglycosidases

Lysosomal exoglycosidases include GAL, GLU, FUC, HEX, as well as MAN. Among lysosomal exoglycosidases, the most active is **HEX** [35] that releases N-acetyloglucosamine and

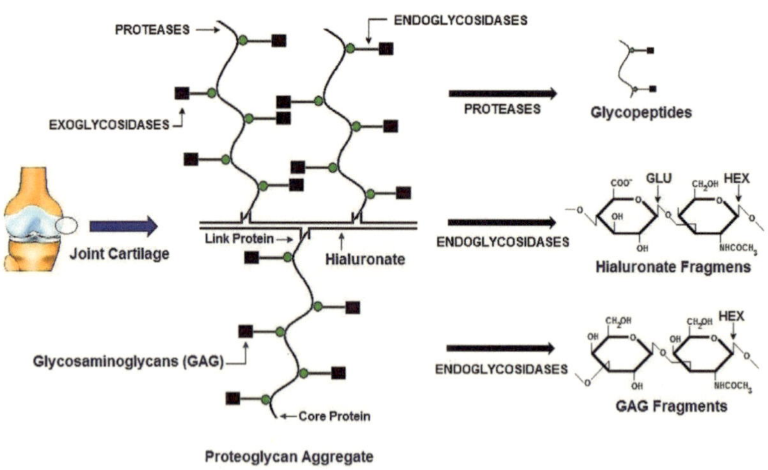

Figure 2. Cartilage destruction by proteases and glycosidases based on [34]. **GAG**—glycosaminoglycans, **GAL**—β-galactosidase, **GLU**—β-glucuronidase, and **HEX**—β-hexosaminidase.

N-acetylgalactosamine from nonreducing ends of oligosaccharide chains of glycoconjugates (glycoproteins, glycolipids, and glycosaminoglycans). HEX is a glycoprotein composed of two (α and β) subunits. HEX subunits create three isoenzymes: HEX A ($\alpha\beta$), HEX B ($\beta\beta$), and HEX S ($\alpha\alpha$) [36]. Predominant are HEX A and HEX B present in different proportions in the particular tissue [19]. Higher concentration of HEX A in comparison to HEX B may be connected with selective degradation of glycosaminoglycans, because HEX A releases N-acetylhexosamines from acid oligosaccharide chains. Some glycolipids (e.g., gangliosides GM2 and GA2) also are degraded exclusively by HEX A. HEX A is thermolabile and undergoes inactivation at pH ~5 after 3 h of incubation at 50°C. Such conditions does not change the activity of HEX B significantly and thus thermal inactivation is used for the differential determination of HEX A and HEX B in biological materials [37].

Human FUC is a glycoprotein occurring in different molecular forms. During separation on Sephadex G-200 column, FUC is eluted in two peaks: α-L-fucosidase I and α-L-fucosidase II. Both isoforms of α-L-fucosidase differ in molecular mass, pH optimum, and susceptibility on heat denaturation (both isoforms are thermolabile, but α-L-fucosidase I undergoes thermal inactivation in basic environment) [38].

Human GAL possesses three isoenzymes: A, B, and C (GAL C activity is small). Isoenzymes A and B absorb at 95% on concanavaline A (ConA) column, and at 60% on wheat germ agglutinin (WGA) column. Absorbance on ConA indicates the presence of mannose, and absorbance on WGA indicates the presence of N-acetylgalactosamine and N-acetylglucosamine in oligosaccharide chains of GAL [39].

Human MAN has three isoenzymes: A, B, and B2 that differ in sialic acid content and spatial arrangement of atoms in macromolecules [40]. Human **GLU** is a tetrameric glycoprotein with MW 310–380 kDa. In humans, there are two forms of β-D-glucuronidase: endogenic and exogenic (bacterial GLU derived from intestinal bacteria: *Escherichia coli, Peptostreptococcus, Bacteroides, Clostridia*) [41, 42]. Increase in the activity of bacterial β-D-glucuronidase in gastrointestinal tract increases detoxication by hydrolysis of glucuronides combined with drugs and other toxic substances, secreted with bile [43, 44]. It was reported that the activity of bacterial β-D-glucuronidase in gastrointestinal tract of persons on high meat diet is significantly higher than that in persons on vegetarian diet [45].

2.1.1. Decrease in the activities of lysosomal exoglycosidases

Both deficiency and excessive HEX activity may have clinical significance. Inherited deficiency in **HEX A** (Tay-Sachs disease) causes storage of nondigested ganglioside GM2 in lysosomes of the neural cells. Enlarged lysosomes oppress remaining organelles of neurons that undergo apoptosis [46]. Up to 5–6 months of life, infants with Tay-Sachs disease develop normally but eventually show impairment of vision, hearing, and movement as well as growing mental retardation. Tay-Sachs children usually live up to 4–5 years of age [35]. Tay-Sachs disease may (but much less often) occur in teenagers and adults, generating milder symptoms [47]. Crucial for the diagnosis of Tay-Sachs disease is the significant decrease in HEX A activity in blood serum, leukocytes, and skin fibroblasts. Tay-Sachs disease may be diagnosed during pregnancy by determination of HEX A activity in amniotic fluid [35].

Deficit in **HEX A and HEX B** activity induces Sandhoff's disease that has symptoms similar to Tay-Sachs disease, with additional storage of oligosaccharides in internal organs, mainly liver and spleen. Children with Sandhoff's disease also live up to 4–5years of age [35]. Sandhoff's disease may appear in older children and adolescents as well as adults, but in older people it occurs extremely rarely. Symptoms of Sandhoff's disease in older children, adolescents, and adults are unspecific and include muscular weakness, lack of motoric coordination (ataxia), speech disorders, and mental retardation. For confirmation of Sandhoff's disease, determination of HEX in blood serum, leukocytes, tears, and skin fibroblasts of potential patients is recommended [35].

Absence or deficiency of **α-L-fucosidase** leads to mucopolysacharydosis, called fucosidosis. Fucosidosis is characterized by lysosomal storage of glycoproteins, glycolipids, and oligosaccharides containing fucose residues. Fucosidosis may be confirmed by the decrease in α-L-fucosidase activity, in fibroblasts or leukocytes of suspected persons [35].

Decreased activity of α-L-fucosidase in breast tissue may be a predisposing factor for the appearance of breast cancer, because high levels of cell surface-associated α-L-fucose are related to neoplastic progression [48].

Deficiency of **β-D-galactosidase** results in generalized gangliosidosis with lysosomal storage of keratan sulfate, oligosaccharide chains of glycoproteins, and gangliosides GM1. Generalized gangliosidosis is manifested by mental retardation, liver enlargement, and bone deformation. Absence of GAL leads to Krabbe disease with storage of galactosyloceramide [35]. Deficiency of **α-L-mannosidase** (mannosidosis) is characterized by primary immune deficiency, skeletal abnormalities, facial dysmorphy, and mental retardation [49]. An inherited lack of **β-D-glucuronidase** activity results in metabolic disease called Sly syndrome or mucopolysaccharidosis VII. Sly syndrome may cause generalized edema of fetus before delivery. Fetuses that survive generalized edema before delivery frequently suffer from mental retardation, littleness, thick facial features, as well as liver and spleen enlargement [35].

2.1.2. Increase in lysosomal exoglycosidase activities

Intensive inflammatory processes, for example tonsillitis, usually are accompanied by increase in lysosomal glycoconjugate catabolism [50]. Hashimoto et al. [51] reported that pancreatic inflammation increases autophagy in the pancreatic inflamed cells. During autophagy, there is observed increase in the activity of lysosomal enzymes characteristic to the involved tissue. The most active of lysosomal exoglycosidases is **HEX**. Therefore, HEX activity in tissues and body fluids is particularly sensitive and is a specific diagnostic parameter for diseases proceeding with increased glycoconjugate degradation. HEX in serum and its isoenzyme B (HEX B) in urine became sensitive and specific markers of alcoholic abuse [52, 53]. A significant increase in serum HEX A activity in smokers may be a marker of risk for arteriosclerosis [54], and an increased HEX A in saliva may be a marker of periodontitis in persons addicted to ethanol [28]. In infectious diseases (e.g., Lyme arthritis), a significant increase in the activity of lysosomal exoglycosidases (including HEX) in serum [55] and synovial fluid was observed [30]. In chronic exposition of rats to cadmium, damage to the proximal renal canalicules and significant increase in HEX and HEX A activities in urine of exposed rats were observed [29].

Determination of HEX in neoplastic tissues presents ambiguous results that depend on circumstances [56]. Generally, in cancerous tissue, increase in the activity of hydrolytic enzymes including HEX should be observed. In tissues of benign neoplasm of human salivary gland a significant increase in HEX and its isoenzymes was observed, in comparison to healthy salivary gland [57]. A significant increase in the activity of lysosomal enzymes (including HEX, HEX A, and HEX B) was reported in malignant brain tissue in comparison to brain tissues without neoplastic changes [58]. But also significant decrease in HEX, HEX A, and HEX B activities in renal cancer tissue in comparison to healthy renal tissue was reported, followed by a significant increase in HEX and its isoenzymes in urine of neoplastic patients in comparison to healthy persons [27]. Therefore, determination of urinary HEX and its isoenzymes may be particularly useful in diagnostics of neoplasms derived from renal epithelial cells of proximal contorted canalicules. Activity of urinary HEX and other exoglycosidases may be helpful in the diagnostics of pancreatic [31] and colon [32] cancers. Detection of HEX and its isoenzyme activity in stools may be used in elaboration of screening markers for detection of the colon cancer [59]. Determination of HEX activity in serum and saliva may be used for the diagnostics and control of salivary gland tumors [25]. The activity of lysosomal α-L-fucosidase (FUC) [60] reflects the intensity of degradation the α-L-fucose containing glycoproteins and glycolipids [25]. The activity of β- galactosidase (GAL) reflects intensity of degradation glycoproteins, glycolipids and glycosaminoglycans containing galactose [39] and activity of β-glucuronidase (GLU) reflects intensity of glycosaminoglycans catabolism [61, 62].

Determination of the activities of FUC, GAL, and GLU may be applied for the diagnostics and monitoring of diseases proceeding with an increase in catabolism of oligosaccharide chains containing sugars released by appropriate exoglycosidases [33]. Increase in the activity of **α-L-fucosidase** in patients with liver cirrhosis seems to be a promising marker for detecting small focuses of liver cancer, particularly when currently used markers (α-fetoprotein and des-γ-carboxy-prothrombin) seem to be less useful than it was primarily expected [63–66]. In the case of **β-D-galactosidase**, there are suggestions that increase in serum GAL activity that may be applied for the diagnostics of glandular colon cancer [59] and larynx cancer [67]. Serum **β-D-glucuronidase** may be a useful marker for recurrence of liver inflammation [68] and increase in proteoglycans degradation in diabetes [69]. Increased activity of serum and tissues β-D-glucuronidase was found in joint inflammation, dermatoses, liver diseases, AIDS, and breast, stomach, rectum, and pancreatic cancers [70].

Activities of the lysosomal exoglycosidases in body fluids are good markers of neoplasms, inflammations, and infections. Determination of exoglycosidase activities in tissues may be helpful in establishing pathogenesis and treatment of some diseases, for example, nasal polyps. Nasal polyps are grape-shaped smooth structures, arising from the inflammatory nasal mucous membrane. Nasal polyps bulge to interior of the nose, restricting nasal patency [71]. There are different pathogenesis theories of nasal polyps, however, none was satisfactorily confirmed, and the lack of understanding nasal polyp pathogenesis impedes therapy. It is known that untreated nasal polyps may cause intra- and extracranial complications. Currently used pharmacological and surgical treatments of nasal polyps do not provide satisfactory results [72]. In nasal polyp tissue, a significant decrease in the concentration of activities of particular exoglycosidases was found in comparison to control, with simultaneous increase in specific activity of HEX A [73, 74]. A decrease in concentration of lysosomal exoglycosidases in nasal polyp tissue, without significant changes in their specific activities, denies the theory of full symptomatic inflammation in nasal polyp pathogenesis and may indicate neoplastic theory.

The activities of lysosomal exoglycosidases may be helpful in the selection of a proper method for treatment of hypertrophied and inflammatory palatal tonsils. Healthy palatal tonsils are important elements of immunological barrier against infections of the respiratory tracts [75, 76]. In the case of hypertrophy of lymphoidal tissue or chronic inflammation of palatal tonsils, otorhinolaryngologists very often face situations where palatal tonsils fail to serve as an immunological barrier and cause complications such as: impeded breathing and swallowing as well as speech disturbance. Palatal tonsils hypertrophy and inflammation are indication for tonsillotomy (trimming) or tonsillectomy (removal of palatal tonsils) [77, 78]. However, some otorhinolaryngologists claim that indications for tonsillo- and tonsillectomy should be limited, especially in younger children (6–7 years old), because the role of palatine tonsils and the possibility of surgical complications are not fully known [77–79]. Popko et al. [80] reported that the activity of lysosomal exoglycosidases in palate tonsils is independent of patients' age and she concluded that probably chronic inflammatory processes of the connective tissue of palate tonsils have the same intensity in childhood and in mature persons, and therefore she recommend tonsillectomy even in childhood.

3. Preparation of tissues and body fluids for determination of lysosomal exoglycosidases

Tissues for exoglycosidase determination were collected during surgery, rinsed in tap water, and then in 0.9% saline. After drying with sterile swab, the tissue was weighed, suspended in 0.154-M KCl with 0.2% Triton X-100 (9 mL of fluid for 1 g of tissue) and homogenized. The homogenate was centrifuged at 40C for 20 min at 12.000 xg. The supernatants were used to determine exoglycosidase activity [25, 72].

Cell culture: Isolated tissue was cut into small pieces (about 0.5 mm^2) and incubated in a mixture prepared in the following proportions: 1.5 mg of collagenase (*Clostridium histolyticum* type I-A) and 1 mg of hyaluronidase in 1 mL of DMEM (Dulbecco's modified Eagle's medium), for 1–2 hours at a temperature of 37°C. After incubation the cells were centrifuged and washed with medium (DMEM), the rinsed cells were then cultivated in plastic bottles (25-cm^2 culture surface) on DMEM with: 10% of calf serum in 25-mM Hepes buffer, 10,000-U/mL penicillin, and 1-mg/mL streptomycin, at 37°C and an atmosphere saturated with 5% CO_2 for –3–5 days, when cells grow up to ~80% of confluence [80].

Articular fluid was collected with sterile syringe needle during local anesthesia, usually at USG control. Collected articular fluid was centrifuged, and exoglycosidases were determined in the supernatant [62].

Urine collection from midstream (after night and after hygiene of intimate places) for determination of exoglycosidases was performed at the same way as for general examination of urine. Exoglycosidase activity determination was performed in supernatants from centrifuged urine [32].

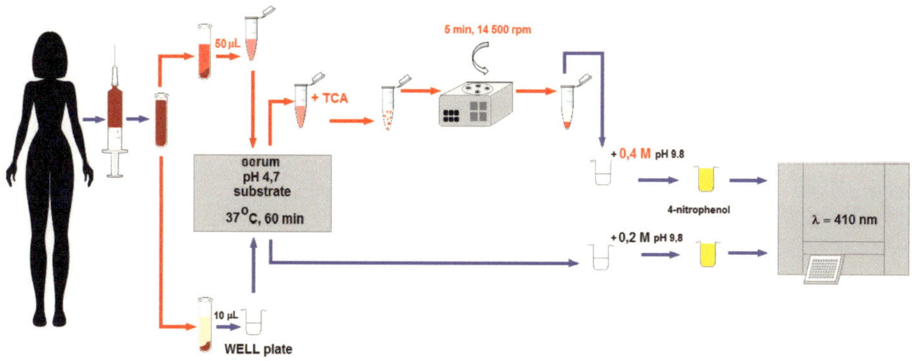

Figure 3. (HEX) determination in the serum of hemolyzed (upper part) and nonhemolyzed blood (lower part). Reprinted from [82], with permission from Elsevier: order number: 4042581223257. **TCA**—trichloroacetic acid.

Saliva secreted to the bottom of the oral cavity was spitted off to the plastic containers placed in crushed ice and then centrifuged; salivary exoglycosidases were determined in the supernatant [74].

Plasma: Blood was collected typically to anticoagulant from cubital vein and then centrifuged for 5 min at 4000 rev/min. Plasma was collected in plastic tubes [81].

Serum: Blood was taken typically from cubital vein, without an anticoagulant. After clotting in room temperature (30–60 min), the clot was centrifuged off at 4000 rev/min for 5 min. Serum was collected in plastic tubes [82].

Tissues immediately after resection, rinsing, and drying were frozen and stored at −80°C for a very long time. In homogenates and supernatant fluids, exoglycosidases should be determined without delay. Synovial fluids, urine, saliva, plasma, and serum may be stored at −80°C.

Determination of lysosomal exoglycosidases is usually performed by modifications of the Chatterjee et al. [83] method based on incubation of suitable tissue homogenates or body fluids with 4-nitro-derivatives of adequate sugar as substrates, in buffered incubation mixture of appropriate pH (4.7 for HEX, 4.5 for GLU, and 4.3 for GAL, FUC, and MAN), at optimal temperature (36°C), and optimal incubation time (60 min). Enzymatic reaction is terminated by alkalization of the incubation mixture (usually by borate buffer at pH 9.8). Chatterjee et al. [83] method was adopted for the determination of exoglycosidases in synovial fluid [62] and saliva [74]. Recently we have published the method for determination HEX [82] in serum from hemolyzed blood (**Figure 3)** [84].

4. Conclusion

The above literature review indicates the activities of lysosomal exoglycosidases in tissues and body fluids as the markers for detection and monitoring of many human diseases.

Author details

Sylwia Chojnowska[1]*, Alina Kępka[2], Slawomir Dariusz Szajda[3], Napoleon Waszkiewicz[3] and Krzysztof Zwierz[1]

*Address all correspondence to: schojnowska@pwsip.edu.pl

1 Medical Institute, Lomza State University of Applied Sciences, Lomza, Poland

2 Department of Biochemistry, Radioimmunology and Experimental Medicine, The Children Memorial Health Institute, Warsaw, Poland

3 Department of Psychiatry, Medical University of Bialystok, Poland

References

[1] Alberts B, Johnson A, Lewis J, Raff M, Roberts K, Walter P. Molecular Biology of the Cell. 5th ed. New York: Garland Science, Taylor & Francis group LLC; 2008. p. 780. ISBN: 978-0815341055

[2] Cooper GM, Sunderland MA. The Cell: A Molecular Approach. 2nd ed. Sinauer Associates, Boston University; 2000. ISBN-10: 0-87893-106-6

[3] Cuervo AM, Dice JF. Lysosomes, a meeting point of proteins, chaperones and proteases. Journal of Molecular Medicine. 1998;**76**:6-12

[4] Lübke T, Lobel P, Sleat DE. Proteomics of the lysosome. Biochimica et Biophysica Acta. 2009;**1793**:625-635. DOI: 10.1016/j.bbamcr.2008.09.018

[5] Schulze H, Kolter T, Sandhoff K. Principles of lysosomal membrane degradation: Cellular topology and biochemistry of lysosomal lipid degradation. Biochimica et Biophysica Acta. 2009;**1793**:674-683. DOI: 10.1016/j.bbamcr.2008.09.020

[6] Alberts B, Johnson A, Lewis J, Raff M, Roberts K, Walter P. Molecular Biology of the Cell. 5th ed. New York: Garland Science, Taylor & Francis group LLC; 2008. pp. 787-789. ISBN: 978-0815341055

[7] Ciechanover A. Intracellular protein degradation: From a vague idea through the lysosome and the ubiquitin proteasome system and onto human diseases and drug targeting. Hematology/the Education Program of the American Society of Hematology. 2006;**1**(12):505-506. DOI: 10.1182/asheducation-2006.1.1

[8] Choi AMK, Ryter SW, Levine B. Autophagy in human health and disease. New England Journal of Medicine. 2013;**368**:651-662. DOI: 10.1056/NEJMra1205406

[9] Li WW, Li J, Bao JK. Microautophagy: Lesser-known self-eating. Cellular and Molecular Life Sciences. 2012;**69**:1125-1136. DOI: 10.1007/s00018-011-0865-5

[10] Cuervo AM, Wong E. Chaperone-mediated autophagy: Roles in disease and aging. Cell Research. 2014;**24**:92-104. DOI: 10.1038/cr.2013.153

[11] Niedźwiedzka-Rystwej P, Deptuła W. Autophagy—an important immunological phenomenon. Postępy Biologii Komórki. 2009;**36**:455-464. DOI: 10.2478/pjvs-2013-0026

[12] Mizushima N, Komatsu M. Autophagy: Renovation of cells and tissues. Cell. 2011;**147**: 728-741. DOI: 10.1016/j.cell.2011.10.026

[13] Levine B, Kroemer G. Autophagy in aging, disease and death: The true identity of a cell death impostor. Cell Death & Differentiation. 2009;**16**:1-2. DOI: 10.1038/cdd.2008.139

[14] Droga-Mazovec G, Bojic L, Petelin A, Ivanova S, Romih R, Repnik U, Salvesen GS, Stoka V, Turk V, Turk B. Cysteine cathepsins trigger caspase-dependent cell death through cleavage of bid and antiapoptotic bcl-2 homologue. Journal of Biological Chemistry. 2008;**283**:19140-19150. DOI: 10.1074/jbc.M802513200

[15] Gorodkiewicz E. Surface plasmon resonance imaging sensor for cathepsin determination based on immobilized cystatin. Protein and Peptide Letters. 2009;**16**:1379-1385. DOI: 10.2174/092986609789353754

[16] Uchiyama Y. Autophagic cell death and its execution by lysosomal cathepsins. Archives of Histology and Cytology. 2001;**164**:233-246

[17] Rossi A, Deveraux Q, Turk B, Sali A. Comprehensive search for cysteine cathepsins in the human genome. Biological Chemistry. 2004;**385**:363-372. DOI: 10.1515/BC.2004.040

[18] Winchester B. Lysosomal metabolism of glycoproteins. Glycobiology. 2005;**15**:1R–15R. DOI: 10.1093/glycob/cwi041

[19] Ostrowska L, Zwierz K, Koniusz Z, Gindzieński A. Role, properties and clinical importance of N-acetyl-β-hexosaminidase. Postępy Higieny i Medycyny Doświadczalnej. 1993;**47**(1):67-79

[20] Levine B, Kroemer G. Autophagy in the pathogenesis of disease. Cell. 2008;**132**:27-42. DOI: 10.1016/j.cell.2007.12.018

[21] Nixon RA, Yang DS, Lee JH. Neurodegenerative lysosomal disorders: A continuum from development to late age. Autophagy. 2008;**4**:590-599

[22] Kirkegaard T, Jäättelä M. Lysosomal involvement in cell death and cancer. Biochimica et Biophysica Acta. 2009;**1793**(4):746-754. DOI: 10.1016/j.bbamcr.2008.09.008

[23] Otomo A, Pan L, Hadano S. Dysregulation of the autophagy-endolysosomal system in amyotrophic lateral sclerosis and related motor neuron diseases. Neurology Research International. 2012;**2012**:498428. DOI: 10.1155/2012/498428

[24] Waszkiewicz N, Zalewska-Szajda B, Chojnowska S, Kępka A, Zasadowska W, Kołodziejczyk P, Dadan J, Zwierz K. Isoenzymes A and B of N-acetyl-β-D-hexosaminidase in tissue of colon cancer. Przegląd Gastroenterologiczny. 2012;**7**(6):374-378. DOI: 10.5114/pg.2012.33045

[25] Bierć M, Minarowski L, Woźniak L, Chojnowska S, Knaś M, Szajda SD, Zwierz K. The activity of selected glycosidases in salivary gland tumors. Folia Histochemica et Cytobiologica. 2010;**48**(3):471-474. DOI: 10.2478/v10042-010-0080-5

[26] Chojnowska S, Minarowska A, Waszkiewicz N, Kępka A, Zalewska-Szajda B, Gościk E, Kowal K, Olszewska E, Konarzewska-Duchnowska E, Minarowski Ł, Zwierz K, Ładny JR, Szajda SD. The activity of N-acetyl-β-D-hexosaminidase A and B and β-glucuronidase in nasal polyps and hypertrophic nasal concha. Otolaryngologia Polska. 2014;**68**(1):20-24. DOI: 10.1016/j.otpol.2013.06.005

[27] Borzym-Kluczyk M, Olszewska E, Szajda SD, Knaś M, Darewicz B, Zwierz K. Activity of isoenzymes A and B of N-acetyl-beta-glucosaminidase in renal cancer tissue. Contemporary Oncology. 2006;**10**(77):502-505

[28] Waszkiewicz N, Zalewska-Szajda B, Chojnowska S, Szajda SD, Zalewska A, Konarzewska B, Szulc A, Wojtulewicz-Supron A, Kępka A, Knaś M, Ładny JR, Milewski R, Zwierz K. The salivary β-HEX A% index as an excellent marker of periodontitis in smoking alcohol-dependent persons. Disease Markers. 2013;**35**(5):457-463. DOI: 10.1155/2013/575074

[29] Brzózka MM, Stypułkowska A, Zwierz K, Moniuszko-Jakoniuk J. Urinary activities of N-acetyl-β-D-glucosaminidase and its isoenzyme B in cadmium-exposed rats. Polish Journal of Environmental Studies. 2004;**13**:121-125

[30] Pancewicz S, Popko J, Rutkowski R, Knaś M, Grygorczuk S, Guszczyn T, Bruczko M, Szajda SD, Zajkowska J, Kondrasiuk M, Sierakowski S, Zwierz K. Activity of lysosomal exoglycosidases in serum and synovial fluid in patients with chronic Lyme and rheumatoid arthritis. Scandinavian Journal of Infectious Diseases. 2009;**41**(8):584-589. DOI: 10.1080/00365540903036220

[31] Szajda SD, Snarska J, Jankowska A, Puchalski Z, Zwierz K. Isoenzymes A and B of N-acetyl-β-D-hexosaminidase in serum and urine of patients with pancreatic cancer. Hepatogastroenterology. 2008;**55**(82-83):695-698

[32] Szajda SD, Snarska J, Puchalski Z, Zwierz K. Lysosomal exoglycosidases in serum and urine of patients with colon adenocarcinoma. Hepatogastroenterology. 2008;**55**(84):921-925

[33] Chojnowska S, Kępka A, Szajda SD, Waszkiewicz N, Bierć M, Zwierz K. Exoglycosidase markers of diseases. Biochemical Society Transactions. 2011;**39**(1):406-409. DOI: 10.1042/BST0390406

[34] Popko J., Zwierz K., Olszewski S., Guszczyn T. Lysosomal Exoglycosidases in Degradation of Human Articular Cartilage. In: Andrew B. Lemmey, editor. Rheumatoid Arthritis – Etiology, Consequences and Co-Morbidities. InTech Open Access Publisher, Rijeka, Croatia; 2012. pp. 97-110. ISBN : 978-953-307-847-2

[35] Vellodi A. Lysosomal storage disorders. British Journal of Haematology. 2005;**128**(4):413-431. DOI: 10.1111/j.1365-2141.2004.05293.x

[36] Mahuran DJ. Characterization of human placental beta-hexosaminidase I 2. Proteolytic processing intermediates of hexosaminidase A. Journal of Biological Chemistry. 1990; **265**(12):6794-6799

[37] Zwierz K, Zalewska A, Zoch-Zwierz W. Isoenzymes of N-acetyl-beta-hexosaminidase. Acta Biochimica Polonica. 1999;**46**(3):739-757

[38] Throppe R, Robinson D. Isoelectric focusing of isoenzymes of human liver alpha-fucosidase. FEBS Letters. 1975;**54**(1):89-92

[39] Czartoryska B. Lysosomal glycosidases in heteropolysaccharide catabolism. Postepy Biochemii. 1977;**23**(2):229-266

[40] Cheng SH, Malcolm S, Pemble S, Winchester B. Purification and comparison of the structure of human liver acidic alpha-mannosidases A and B. Biochemical Journal. 1986; **233**(1):65-72

[41] Dohrmann RE. β-Glucuronidase. Berlin, Heidelberg, New York: Springer Verlag; 1969. ISBN: 978-3-642-95089-6

[42] Levvy GA, Conchie J. β-glucuronidase and the hydrolysis of glucuronides. In: Dutton GJ, editor. Glucuronic Acid: Free and Combined. New York: Academic Press; 1966. pp. 301-364. ISBN: 978-0-12-395501-2

[43] Hanausek M, Walaszek Z, Slaga TJ. Detoxifying cancer causing agents to prevent cancer. Integrative Cancer Therapies. 2003;**2**(2):139-144. DOI: 10.1177/1534735403002002005

[44] Walaszek Z. Chemopreventive properties of D-glucaric acid derivatives. Cancer Bulletin. 1993;**45**:453-457

[45] Żółtaszek R, Hanausek M, Kiliańska ZM, Walaszek Z. The biological role of D-glucaric acid and its derivatives: Potential use in medicine. Postępy Higieny i Medycyny Doświadczalnej. 2008;**62**:451-462

[46] Mahuran DJ. Biochemical consequences of mutations causing the GM2 gangliosidoses. Biochimica et Biophysica Acta. 1999;**1455**(2-3):105-138

[47] Kaback MM, Desnick RJ. Hexosaminidase A deficiency. In: Pagon RA, Adam MP, Ardinger HH, Wallace SE, Amemiya A, Bean LJH, Bird TD, Fong CT, Mefford HC, Smith RJH, Stephens K, editors. GeneReviews® [Internet]. Seattle, WA: University of Washington, Seattle; 1993-2016

[48] Cheng TC, Tu SH, Chen LC, Chen MY, Chen WY, Lin YK, Ho CT, Lin SY, Wu CH, Ho YS. Down-regulation of α-L-fucosidase 1 expression confers inferior survival for triple-negative breast cancer patients by modulating the glycosylation status of the tumor cell surface. Oncotarget. 2015;**6**(25):21283-21300. DOI: 10.18632/oncotarget.4238

[49] Malm D, Nilssen O. Alpha-mannosidosis. Orphanet Journal of Rare Diseases. 2008;**3**(1):21. DOI: 10.1186/1750-1172-3-21

[50] Popko M. Lysosomal exoglycosidases in human palatine tonsils [thesis]. Białystok: Medical University of Bialystok; 2008

[51] Hashimoto D, Ohmuraya M, Hirota M, Yamamoto A, Suyama K, Ida S, Okumura Y, Takahashi E, Kido H, Araki K, Baba H, Mizushima N, Yamamura K. Involvement of autophagy in trypsinogen activation within the pancreatic acinar cells. Journal of Cell Biology. 2008;**181**(7):1065-1072. DOI: 10.1083/jcb.200712156

[52] Hultberg B, Isaksson A, Berglund M, Moberg AL. Serum β-hexosaminidase isoenzyme: A sensitive marker of alcohol abuse. Alcoholism, Clinical and Experimental Research. 1991;**15**(3):549-552

[53] Waszkiewicz N, Popławska R, Konarzewska B, Szajda SD, Galińska B, Rutkowski P, Leśniak R, Szulc A. Biomarkers of alcohol abuse. Part II. New biomarkers and their interpretation. Psychiatria Polska. 2010;**44**(1):137-146

[54] Hultberg B, Isaksson A, Nilsson JA, Lindarde F. Serum β-hexosaminidase isoenzymes are related to risk factors for atherosclerosis in a large population of post menopausal women. Clinica Chimica Acta. 1994;**227**(1-2):59-68

[55] Pancewicz SA, Wielgat P, Hermanowska-Szpakowicz T, Kondrusiuk M, Zajkowska J, Grygorczuk A, Popko J, Zwierz K. Activity of lysosomal exoglycosidases in the serum of patients with chronic Lyme arthritis. International Journal of Medical Microbiology. 2006;**296**(40):280-282. DOI: 10.1016/j.ijmm.2006.01.002

[56] Brech A, Ahlquist T, Lothe RA, Stenmark H. Autophagy in tumour suppression and promotion. Molecular Oncology. 2009;**3**(4):366-375. DOI: 10.1016/j.molonc.2009.05.007

[57] Borzym-Kluczyk M, Olszewska E, Radziejewska I, Lewszuk A, Zwierz K. Isoenzymes of N-acetyl-beta-hexosaminidase in human pleomorphic adenoma and healthy salivary glands: A preliminary study. Clinical Chemistry and Laboratory Medicine. 2008;**46**(1):131-136. DOI: 10.1515/CCLM.2008.018

[58] Wielgat B, Walczuk U, Szajda S, Bień M, Zimnoch I, Mariak Z, Zwierz K. Activity of lysosomal exoglycosidases in human gliomas. Journal of Neuro-Oncology. 2006;**80**(3):243-249. DOI: 10.1007/s11060-006-9188-z

[59] Waszkiewicz N, Szajda SD, Konarzewska-Duchnowska E, Zalewska-Szajda B, Gałązkowski R, Sawko A, Nammous H, Buko V, Szulc A, Zwierz K, Ładny JR. Serum β-glucuronidase as a potential colon cancer marker: A preliminary study. Postępy Higieny i Medycyny Doświadczalnej. 2015;**69**:436-439

[60] Li C, Qian J, Lin JS. Purification and characterization of alpha-L-fucosidase from human primary hepatocarcinoma tissue. World Journal of Gastroenterology. 2006;**12**(23): 3770-3775

[61] Buddecke E. Proteoglycans. In: Gabius HJ, editor. The Sugar Code - Fundamentals of Glycosciences. Wiley-VCH Verlag GmbH & Co. KGaA; Weinheim, Federal Republic of Germany, 2009. pp. 199-216. ISBN:978-3-527-32089-9

[62] Marciniak J, Zalewska A, Popko J, Zwierz K. Optimization of an enzymatic method for the determination of lysosomal N-acetyl-beta-D-hexosaminidase and beta glucuronidase in synovial fluid. Clinical Chemistry and Laboratory Medicine. 2006;**44**(8):933-937. DOI: 10.1515/CCLM.2006.177

[63] Takahashi H, Saibara T, Iwamura S, Tomita A, Maeda T, Onishi S, Yamamoto Y, Enzan H. Serum alpha-L-fucosidase activity and tumor size in hepatocellular carcinoma. Hepatology. 1994;**19**(6):1414-1417

[64] Giardina MG, Matarazzo M, Morante R, Lucariello A, Varriale A, Guardasole V, De Marco G. Serum α-L-fucosidase activity and early detection of hepatocellular carcinoma. Cancer. 1998;**83**(12):2468-2474

[65] Gan Y, Liang Q, Song X. Diagnostic value of alpha-L-fucosidase for hepatocellular carcinoma: A meta-analysis. Tumour Biology. 2014;**35**(5):3953-3960. DOI: 10.1007/s13277-013-1563-8

[66] Wang K, Guo W, Li N, Shi J, Zhang C, Lau WY, Wu M, Cheng S. Alpha-L-fucosidase as a prognostic indicator for hepatocellular carcinoma following hepatotectomy: A large-scale, long-term study. British Journal of Cancer. 2014;**110**(7):1811-1819. DOI: 10.1038/bjc.2014.102

[67] Olszewska E, Borzym-Kluczyk M, Rzewnicki I, Wojtowicz J, Rogowski M, Pietruski JK, Czajkowska A, Sieśkiewicz A. Possible role of α-mannosidase and β-galactosidase in larynx cancer. Contemporary Oncology. 2012;**16**(2):154-158. DOI: 10.5114/wo.2012.28795

[68] Ohta H. Measurement of serum immunoreactive beta-glucuronidase: A possible serological marker for histological hepatic cell necrosis and to predict the histological progression of hepatitis. Hokkaido Igaku Zasshi. 1991;**66**(4):45-57

[69] Miller MD, Keyes FP, Curreri PW. Increase of serum β-glucuronidase activity in human diabetes mellitus. The Journal of the American Medical Association. 1966;**195**(3):189-192

[70] Severini G, Diana L, Di Giovannandrea R, Tirelli C. A study of serum glycosidases in cancer. Journal of Cancer Research and Clinical Oncology. 1995;**121**(1):61-63

[71] Settipane RA, Peters AT, Chiu AG. Nasal polyps. American Journal of Rhinology & Allergy. 2013;**27**(1):S20–S25. DOI: 10.2500/ajra.2013.27.3926

[72] Chojnowska S. Lysosomal exoglycosidases in nasal polyps [thesis]. Białystok: Medical University of Bialystok; 2009

[73] Chojnowska S, Minarowska A, Knaś M, Niemcunowicz-Janica A, Kołodziejczyk P, Zalewska-Szajda B, Kępka A, Minarowski Ł, Waszkiewicz N, Zwierz K, Szajda SD. Lysosomal exoglycosidases in nasal polyps. Otolaryngologia Polska. 2013;**67**(4):192-197. DOI: 10.1016/j.otpol.2013.05.004

[74] Chojnowska S, Zalewska A, Knaś M, Waszkiewnicz N, Waszkiel D, Kossakowska A, Zwierz K. Determination of lysosomal exoglycosidases in human saliva. Acta Biochimica Polonica. 2014;**61**(1):85-90

[75] Brandtzaeg P, Halstensen TS. Immunology and immunopathology of tonsils. Advances in Oto-Rhino-Laryngology. 1992;**47**:64-75

[76] Ikinciogullari A. Is immune system influenced by adenotonsillectomy in children? International Journal of Pediatric Otorhinolaryngology. 2002;**66**(3):251-257

[77] Brodsky L, Poje C. Tonsillitis, tonsillectomy, and adenoidectomy. In: Bailey BJ, Johnson JT, editors. Head & Neck Surgery-Otolaryngology. 5th ed. Philadelphia: Lippincott Williams & Wilkins; 2006. pp. 1184-1197. ISBN-13: 978-0781755610

[78] Paulussen C. Adenoids and tonsils, indications for surgery and immunological consequences of surgery. Acta Oto-Rhino-Laryngologica Belgica. 2000;**54**(3):403-408

[79] Zhang PC, Pang YT, Loh KS, Wang DY. Comparison of histology between recurrent tonsillitis and tonsillar hypertrophy. Clinical Otolaryngology and Allied Sciences. 2003;**28**(3):235-239

[80] Popko J, Marciniak J, Ilendo E, Knaś M, Guszczyn T, Stasiak-Barmuta A, Moniuszko T, Zwierz K, Wysocka J. Profile of exoglycosidases in synovial cell cultures derived from human synovial membrane. Cell Biochemistry and Biophysics. 2008;**51**(2-3):89-95. DOI: 10.1007/s12013-008-9018-3

[81] Kępka A, Waszkiewicz N, Zalewska-Szajda B, Chojnowska S, Płudowski P, Konarzewska E, Szulc A, Ładny JR, Zwierz K, Szajda SD. Plasma carnitine concentration after chronic alcohol intoxication. Postępy Higieny i Medycyny Doświadczalnej. 2013;**31**:548-552

[82] Chojnowska S, Kępka A, Szajda SD, Kołodziejczyk P, Zwierz K, Waszkiewicz N. Determination of N-acetyl-β-hexosaminidase in serum from hemolyzed blood. Clinical Biochemistry. 2016;**49**(10-11):811-815. DOI: 10.1016/j.clinbiochem.2016.03.002

[83] Chatterjee S, Velicer LF, Sweeley CC. Glycosphingolipid glycosyl hydrolases and glycosidases of synchronized human KB cells. Journal of Biological Chemistry. 1975;**250**: 4972-4979

[84] Chojnowska S, Kępka A, Szajda SD, Waszkiewicz N, Zwierz K. Determination of lysosomal exoglycosidases in hemolyzed blood. Acta Biochimica Polonica. 2016;**63**(2):202